E. W. (Edward William) Young

Simple, Practical Methods of Calculating Strains on Girders, Arches,

and Trusses

E. W. (Edward William) Young

Simple, Practical Methods of Calculating Strains on Girders, Arches, and Trusses

ISBN/EAN: 9783744661867

Printed in Europe, USA, Canada, Australia, Japan

Cover: Foto ©berggeist007 / pixelio.de

More available books at **www.hansebooks.com**

SIMPLE PRACTICAL METHODS

OF CALCULATING

STRAINS ON GIRDERS, ARCHES, AND TRUSSES;

WITH A SUPPLEMENTARY ESSAY ON

ECONOMY IN SUSPENSION BRIDGES.

BY

E. W. YOUNG,

ASSOCIATE OF KING'S COLLEGE, LONDON,
AND MEMBER OF THE INSTITUTION OF CIVIL ENGINEERS.

London:
MACMILLAN AND CO.
1873

[All Rights reserved.]

Cambridge:
PRINTED BY C. J. CLAY, M.A.
AT THE UNIVERSITY PRESS.

PREFACE.

THE bulk of the following work was written several years ago, and was designed to supply a want which was then much felt, that of a treatise giving simple methods of calculating strains on girders, trusses, arches, roofs, and other kindred structures. The treatises existing at that time were either too meagre or too mathematical for the average engineering student. One class gave him results without explaining to him how they were obtained; the other, in making the required explanation, bewildered him with mathematical formulæ beyond his comprehension.

Several excellent works on this subject having appeared during the last few years, the issue of another from the press might be thought superfluous. I have judged otherwise, because the methods adopted in this little work are mostly original, and, as I consider, simpler than usual. A knowledge of simple equations, excepting on the subject of deflection, where a little elementary trigonometry is required, is the extent of the mathematical attainment required for working out the problems presented in this work to the reader.

The article on "Economy in Suspension Bridges" from "Engineering" is added as a supplement to the chapter on Suspension Bridges.

THE AUTHOR.

CONTENTS.

PAGE

PREFACE ... iii

CHAPTER I.

OF LEVERS AND THE RESOLUTION OF FORCES.

The Lever. Fulcrum. Arm or leverage. Moment. Levers of three kinds. Bent Lever. Examples. Parallelogram of Forces. Resultant. Triangle of Forces. Corollaries. Polygon of Forces. Examples of the Polygon of Forces. Resolution or Decomposition of Forces. Example. Alternative methods, the Lever or Parallelogram of Forces. Lever method. Method of Parallelogram of Forces. Forces acting in different planes 1

CHAPTER II.

GIRDERS WITH PARALLEL FLANGES.

Load at centre of span. Load evenly distributed, strain at centre. Strain at any point of the Flange. Example. Irregularly loaded Girders. Strain on the web. Shearing force. Cantilevers. Load at the extremity. Distributed load. Wooden beams. Neutral axis. The web in Iron Girders. Examples. Strains on the web. Result checked by method described, p. 14. Example. Formula for Shearing Force. Irregular loading, First case, Bays equal. Second case, Bays of an unequal size, Load symmetrical. Third case, Bays and Loading irregular. Trusses and Lattice Girders. Without verticals. Warren Truss. Strains on Diagonals 12

CHAPTER III.

HOGBACKED GIRDERS.

Strains on the Flanges. Strains on the Diagonals. Comparison with a parallel Flanged Girder. Diagonals as struts. Diagonals of both kinds used 41

CHAPTER IV.

THE BOWSTRING GIRDER.

Strains on the Flanges. Curve of Equilibrium. Method of drawing the Curve of Equilibrium. Regular Loading. To draw the Curve of Equilibrium with a given versine. Irregular Loading, Bays equal. Irregular Loading, Bays unequal. Line of greatest Resistance. Methods of resisting Distortion. First Method. Second Method. Third Method. Fourth Method. Fifth Method. Effect of elasticity in distributing the Load among the Diagonals. Load resting on the Top 45

CHAPTER V.

THE ARCH.

Method of finding the Thrust at Crown. Curve of Equilibrium for an Arch of Masonry. Other methods of resisting Distortion of the Arch. Strains on the Spandrils. Other forms of Spandril filling. Arched bridges of multiple span. Stability of Pier. Effect of Horizontal Girder. Transverse strain on Pier. Stability of a Wrought-iron Arched Viaduct 61

CHAPTER VI.

SUSPENSION BRIDGES.

Strain at centre of Span. Strain at any portion of the Chain. Strain on the Back-tie. Methods of stiffening 72

CHAPTER VII.

DEFLECTION.

Definition. Laws of Deflection. Radius of Curvature. To find the Deflection. Example. Irregularly strained Girder. Deflection of an Arched Rib. Example. Deflection of a Bowstring Girder. Deflection of a Suspension Bridge. Strength of a Girder, how far deducible from its deflection under load 76

CHAPTER VIII.

CONTINUOUS GIRDERS.

Definition. Continuous Girder of two spans. Both spans loaded. One span only loaded. Method of finding position of Point of Contrary Flexure. Example. Two unequal spans. Both spans loaded. Longer span loaded. Extremely disproportioned spans. Shorter span loaded. Anchoring down the shorter Girder. Multiple spans. Three spans equally loaded. Spans unequally loaded. Position of Point of Contrary Flexure in three-span Girder. Girders of more than three spans. Continuous Girders of varying depth. Practical remarks 86

CHAPTER IX.

ROOFING TRUSSES.

Elementary form of Principal. Method of finding the strains. More complicated forms of Principal. Arched Roofs. First System. Second System. Example, Diagonals, both Ties. Third System. Equally loaded. Unequally loaded. Force of Wind. Pressure of Wind on Inclined Planes. Methods of finding strains produced by force of Wind 109

CHAPTER X.

Economy in Suspension Bridges.

PAGE

The question hitherto not sufficiently investigated. Subject considered under two heads. First head: The most economical height for the towers. For load at centre of span. For distributed load. Method of estimating for towers of different kinds of material. Effect of weight of chains not taken into account. Second head: The most economical arrangements of parts in the superstructure. Suspension and cantilever systems compared. For a dead load. For a live load. Proper position for the point of junction of Cantilever and Suspension Systems. Practical part of the question. Bending of platform caused by rolling load. Caused by changes of temperature. Arrangements thereby necessitated. Great economy of the Compound System 119

ERRATA.

Page 33, line 3, *for* F have two *read* F has two.
Page 93, line 8, *for* loading strains *read* loading shewn.
Page 119, line 15, *for* proportion of depth of space *read* proportion of depth to space.

STRAINS ON GIRDERS, ARCHES, AND TRUSSES.

CHAPTER I.

OF LEVERS AND THE RESOLUTION OF FORCES.

The Lever. By "a lever" is commonly understood a bar of wood or iron such as a handspike or crowbar employed for moving great weights. As we shall have under our consideration in the course of this work structures consisting of a combination of many levers, the whole of the structure itself being sometimes considered as one lever, it will be advisable to commence with a few remarks on levers and leverage, a correct knowledge of the principles of which is indispensable for the determination of the strains to which such complex structures are subjected by the pressures which they have to bear.

Fulcrum. The bar abc (fig. 1) is a representation of a simple lever resting on a support at b called the *fulcrum*, or fixed point about which the lever would turn if the forces at each end did not balance each other. A weight of 100 lbs. is suspended from the end c.

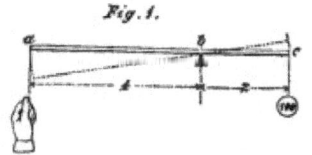

Fig. 1.

Arm or Leverage. Suppose the lever to have no appreciable weight, to be 6 feet long, to lie horizontally, and the fulcrum b to be placed 2 feet from the end c, and therefore 4 feet from a.

Y.   1

The distance bc, being the perpendicular distance of the line of action of the load from the fulcrum, is termed the *Arm* or *Leverage* at which the 100 lb. weight acts. The tendency of the load at c is to depress that extremity of the lever and raise the other. The fulcrum b may be considered as the centre of two concentric circles of which bc and ba are radii; and, assuming the lever to oscillate about the point b, the arcs described by the points a and c and their velocities will be directly proportional to their respective distances from the point b.

Moment. Since it is a first principle that the moments of two opposing forces must be equal in order that they may balance one another, and the moment of any force about a fulcrum is the force multiplied by the leverage, if we divide the moment of the 100 lb. weight by the leverage of the counterbalancing force at a, we shall obtain the value of the latter.

Now the moment of the 100 lb. weight $= 100 \times 2 = 200$;

$\therefore \dfrac{200}{4} = 50$ lbs. $=$ counterbalancing force or *power* acting downwards at a.

The pressure on the fulcrum at b will therefore be 150 lbs.

Levers of three kinds. Figs. 1, 2 and 3 represent what are sometimes called the three kinds of levers, because the relative

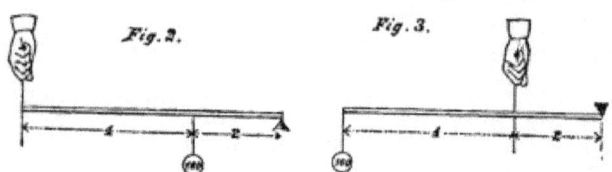

positions of the power, the weight, and the fulcrum are different in each case.

Such a distinction is however unnecessary, and may be confusing to some. We will therefore consider the pressures exerted by the power, the weight, and the fulcrum, simply as *forces*, and represent them by the pull of weights on strings (see fig. 4).

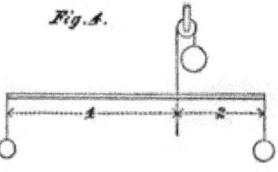

We can then assume at will any of the three points as a fulcrum, and thus avoid confusion of ideas.

Bent Lever. In estimating the moment of the force acting upon the extremity d of the bent lever shewn in fig. 5, the distance bd must not be assumed as the arm or leverage but bc, since bc is the perpendicular distance of the line of action of this force from the fulcrum at b.

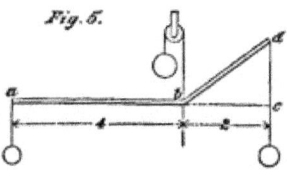

Fig. 5.

Examples. In the example illustrated in fig. 6, the moments of the three forces on the left side of the fulcrum at b are

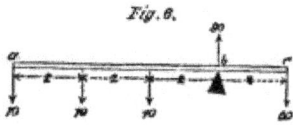

Fig. 6.

$$10 \times 2 = 20$$
$$10 \times 4 = 40$$
$$10 \times 6 = 60$$
$$\overline{120}$$

$\therefore \dfrac{120}{2} = 60$, the force required at c to maintain equilibrium.

The pressure on the fulcrum at $b = 30 + 60 = 90$. The moment of the three forces of 10, each acting at leverages of 2, 4 and 6 respectively, is equal to that of a force of 30 acting at their mean leverage of 4—thus $30 \times 4 = 120$—their moment, and generally—The moment of any number of parallel forces about a fulcrum is equal to the sum of the forces acting at a point which may be called their centre of gravity.

For instance, in fig. 6, the point b is the centre of gravity of the downward forces acting on the lever abc, and their resultant is a pressure of 90, their sum on the fulcrum at that point.

It must be carefully remembered that the moments of forces acting in *contrary directions* must be *added* together when they are situated on *opposite sides* of the fulcrum, because each tends to turn the lever in the same direction.

Fig. 7 is a representation of a lever acted upon by 7 forces, acting either vertically upwards or downwards as indicated by the arrows. Their points of application are equidistant from

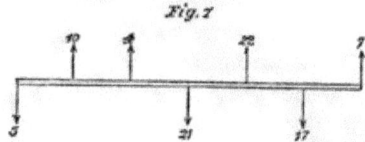
Fig. 1

each other, and their intensities, being represented by numbers written against the arrows, are so arranged as to maintain the lever in equilibrium.

The sum of the upward forces is equal to the sum of the downward forces, and any point in the lever being assumed as a fulcrum, the moments of opposing forces about that point balance each other, e.g.

Let the force 21 represent the reaction of a fulcrum, and let the forces tending to depress the right-hand extremity of the lever have the sign +, and those tending to turn the lever in the opposite direction the sign −, then

Moments of forces on left side of fulcrum are $\begin{cases} -\ 5 \times 3 = -15 \\ +10 \times 2 = +20 \\ +\ 4 \times 1 = +\ 4 \end{cases}$

Total .... + 9

Moments of forces on right side of fulcrum are $\begin{cases} -22 \times 1 = -22 \\ +17 \times 2 = +34 \\ -\ 7 \times 3 = -21 \end{cases}$

Total .... − 9

Again, take the left-hand extremity of the lever as the fulcrum.

The moments are as follows:

$$\begin{array}{rr} -\ 7 \times 6 = -\ 42 & \\ +17 \times 5 = & +\ 85 \\ -22 \times 4 = -\ 88 & \\ +21 \times 3 = & +\ 63 \\ -\ 4 \times 2 = -\ 8 & \\ -10 \times 1 = -\ 10 & \\ \hline \text{Totals} \ldots\ldots\ldots\ldots\ldots -148 & +148 \end{array}$$

Once more, take a point half way between the points of application of the forces 10 and 4 as the fulcrum.

The moments are

$$
\begin{aligned}
+\ 5 \times 1\tfrac{1}{2} &= +\ 7\tfrac{1}{2} \\
-10 \times \tfrac{1}{2} &= & -5 \\
+\ 4 \times \tfrac{1}{2} &= +\ 2 \\
-21 \times 1\tfrac{1}{2} &= & -31\tfrac{1}{2} \\
+22 \times 2\tfrac{1}{2} &= +55 \\
-17 \times 3\tfrac{1}{2} &= & -59\tfrac{1}{2} \\
+\ 7 \times 4\tfrac{1}{2} &= +31\tfrac{1}{2}
\end{aligned}
$$

Totals .... +96   −96

The lever is thus shewn to be in equilibrium.

Fig. 8 is a representation of an irregularly shaped body acted upon by three forces f^1, f^2, and f^3, in the same plane, whose lines of action are shewn by dotted lines, and their directions by arrows.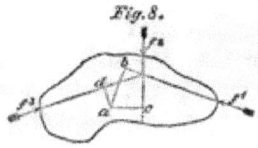
If these forces are in equilibrium, whatever point in the body may be assumed as the fulcrum, the opposing moments of the forces about that point will be found to balance.

Any point a in the body being assumed as the fulcrum the leverages of the forces f^1, f^2, and f^3 may severally be represented by the lines ab, ac, and ad drawn perpendicularly to their respective lines of action from the fulcrum a.

If equilibrium be maintained

$$f^1 \times ab = f^2 \times ac + f^3 \times ad.$$

Although the strains on engineering structures of a very complicated kind can be found by considering them as combinations of levers and applying the principles which have been laid down in the preceding pages, nevertheless the strains in certain cases can be determined more rapidly, and with less liability of error, by the application of the principle known as *The parallelogram or triangle of forces*, which is as follows.

Parallelogram of Forces. *If two adjacent sides of a parallelogram represent both in magnitude and direction two forces meeting in a point, their resultant may be represented by the diagonal of the parallelogram drawn to the point where they meet.*

For example, let *a* (fig. 9) be a body acted upon by two forces represented by the lines *ac* and *ab*, and their directions by the arrows. If *cd** be drawn parallel to *ab*, and *bd* to *ac*, the diagonal *ad* represents both in magnitude and direction a single force equal to the combined effect of the forces *ab* and *ac* upon the body *a*.

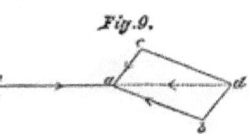

Resultant. The line *ae*, which is the equal of the diagonal *ad*, called the *resultant* of the forces *ab* and *ac*, shews the magnitude and direction of the force necessary to balance the forces *ab* and *ac*.

Triangle of Forces. The forces which keep the body *a* in equilibrium are represented by the sides of the triangle *abd*, both as to direction and amount. The arrows indicating direction arrange themselves as in fig 10, in consecutive order.

Corollary 1. When three forces which meet in a point, and whose directions lie in the same plane, maintain a body in equilibrium; their magnitudes are proportional to the sides of a triangle, each of which is drawn parallel to one of the forces.

Corollary 2. One of the three forces being known, the magnitude of the other two may be found by constructing the triangle.

Corollary 3. Any one force in the triangle is the resultant of the other two.

Polygon of Forces. In fig. 10 substitute for the force *ad* the two forces represented by the dotted lines *ac* and *ad*, of which it may be regarded as the resultant. Also for the force *ab* substitute the forces *be* and *ea*. The condition of equilibrium remains unimpaired.

Thus we have a body maintained in equilibrium by the action of five forces represented by the sides of the polygon *acdbe*.

* This operation is generally called completing the parallelogram.

Hence we may gather that if a body in equilibrium is acted upon by more than three forces meeting in a point, they are proportional to the sides of a polygon to which their directions are severally parallel.

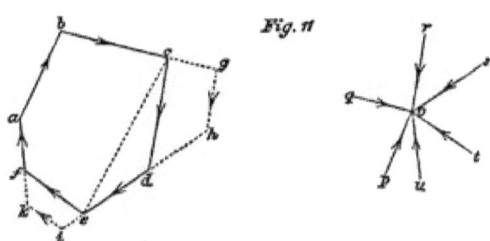

Fig. 11

But since many polygons may be drawn whose homologous sides shall be parallel, it is obvious that we cannot by this means discover the value of unknown forces, except in certain cases.

A body o, fig. 11, is acted upon by 6 forces p, q, r, s, t, u, which may be respectively represented by the sides ab, bc, cd, de, ef, and fa of the polygon $abcdef$, of which ab is drawn parallel to the direction of p, bc to q, and so on, or the forces may be represented by any other polygon as $abcghik$ whose sides are respectively parallel to the forces.

Examples of the Polygon of Forces. To find the values of unknown forces by the polygon of forces.

Of the forces p, q, r, s, t, u, let the values of p, q, r and s be known, and of t and u unknown.

To find t and u.

Draw the line ab parallel to the direction of force p, and make the length of ab to represent the intensity of p. From the point b draw the line bc in a similar manner parallel and equal to force q, cd equal to r, and de equal to s. From the point e draw a line of indefinite length parallel to t, and from point a a line parallel to u intersecting the line drawn from e parallel to t in f; the lines ef and fa represent the values of the forces t and u respectively.

Again, let the forces q and t represented by the sides bc and ef in the polygon be unknown, required their values.

Now as the sides fa, ab, cd and de represent known forces, their lengths and inclination are fixed, consequently the relative position of the points f and b is fixed, also that of c and e. Join ec (fig. 11). Through b (fig. 12) draw a line of indefinite length parallel to the direction of q, also a line through point f, parallel to t; in this line take any point x, draw xy parallel and equal to ec (fig. 11), and through the point y draw a line parallel

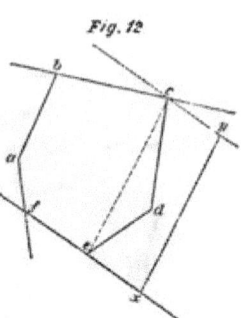

Fig. 12

to fx intersecting the line drawn through b parallel to q in the point c. From c draw a line parallel to the line ce (fig. 11), and meeting the line fx in e; the lines bc and fe represent the values of the two unknown forces.

If the directions of the two unknown forces be *parallel*, their values cannot be discovered.

This method is frequently of use in determining the strains on roofs or other structures where a number of struts and ties meet in a point; the strains on some of them being known, this furnishes a means of arriving at the strains on the others.

Corollary. In the polygon of forces any one force, its direction being reversed, is the resultant of all the other forces.

Thus if the body o be acted upon by 5 forces represented by sides ab, bc, cd, de and ef of the polygon $abcdef$ (fig. 11), their combined effect tends to move the body in the direction of a to f with a force represented by the line af, since the addition of the force fa will keep the body at rest.

Resolution or Decomposition of Forces. In constructing the polygon $acdbe$ (fig. 10), the force ad was *resolved* or *decomposed* into two forces ac and cd. Forces inclined at an angle with the horizontal may be resolved into a horizontal and a vertical force. For example:—the force ac (fig. 13) may be resolved into the horizontal force ab, and the vertical force bc; these forces being termed respectively the horizontal and vertical *elements* or *components* of the force ac.

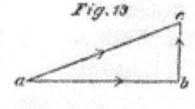

Fig. 13

OF LEVERS AND FORCES.

We shall find it very convenient, in tracing the strains through complicated structures, not to attempt to find the *direct* strain on every bar in the first instance, but the *horizontal* or *vertical* strain merely; from either of these the direct strain can afterwards be calculated.

Example. By the resolution of forces we obtain the following proof that the sides of a polygon represent a set of forces in equilibrium.

Let the sides of the polygon *abcdef* represent six forces acting upon a point, the arrow-heads shewing the directions of the forces which are consecutive.

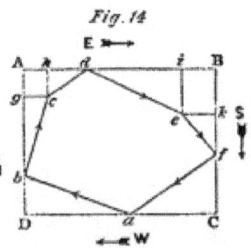

Fig. 14

Describe the rectangle $ABCD$ about the polygon *abcdef*. From c draw cg parallel to AB, and ch parallel to AD, and from e draw ei parallel to BC, and ek parallel to AB.

Let the effect of any force towards the upper part of the parallelogram be called northing, represented by the letter N, that towards the right hand easting, represented by E, and similarly southing and westing by S. and W.

Now the force ab exerts a pressure represented by the line aD in a westerly direction, and in a northerly a pressure equal to Db.

Resolving each force into its N. S. E. or W. directions, and adding all the N.s, S.s, E.s and W.s together, we get the following result.

		N.	S.	E.	W.
Force	$ab =$	Db			aD
„	$bc =$	bg		$(gc =) Ah$	
„	$cd = (ch =)$	gA		hd	
„	$de =$		$(ie =) Bk$	di	
„	$ef =$		kf	$(ek =) iB$	
„	$fa =$		fC		Ca
Sums are equal to		DA	BC	AB	CD

But the forces DA and BC are equal and opposite.
Also the forces AB and CD are equal and opposite.

Therefore they are in equilibrium, and as the combined effect of the forces ab, bc, cd, de, ef, fa has been shewn to be equal to that of the forces DA, BC, AB and CD, therefore the forces represented by the sides of the polygon $abcdef$ are in equilibrium.

Alternative methods, the Lever or Parallelogram of Forces.
As it has been already stated, the strains upon the parts of a structure may be found by means of the parallelogram of forces as well as by the principles of the lever.

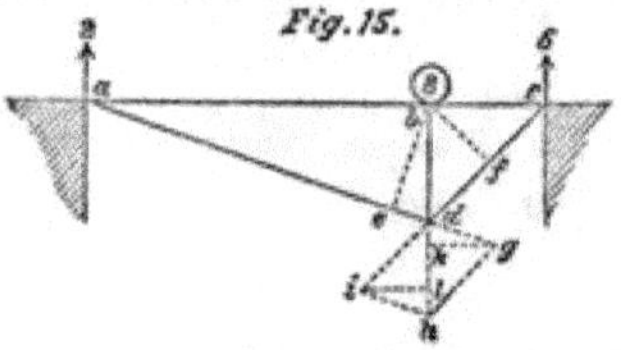

Fig. 15.

$abcd$ (fig. 15) is a simple truss, in which ac is horizontal and bd vertical, supporting a load of 8 at the point b. The depth $bd = \dfrac{ac}{4}$, and the distance $bc = \dfrac{ac}{4}$. Then the load on the point of support at a will be 2 and that on c will be 6.

Lever method. Taking the distance bc as our unit of length, we have 6 as the moment of each of the reactionary pressures of the abutments. Assuming the point b as the fulcrum, 6 being the moment of the force acting at a, the balancing strain on the bar ad must $=\dfrac{6}{eb}$, the distance eb being its leverage. Similarly the strain on the bar cd must $=\dfrac{6}{bf}$.

Again, taking point d as the fulcrum, we find the strain on the bars ab and bc to be $\dfrac{6}{bd} = 6$, since $bd = bc = 1$.

Since $eb : bd :: ab : ad$, we perceive that the strains on the bars ab and ad are to each other as their lengths, and the same holds good of the bars bc and cd.

Method of Parallelogram of Forces. To find the strains by the parallelogram of forces, produce the line bd to h making $dh =$ the load of 8 to any convenient scale. Produce the lines ad and cd indefinitely. From the point h draw hg parallel to cd, and hi parallel to ad. From the points i and g draw horizontal lines il and gk to meet the line dh in the points l and k.

The line dh being the diagonal of the parallelogram $dghi$ represents a force which is the resultant of two forces represented by the lines di and dg, wherefore by applying the same scale to these lines as we have used for dh, we ascertain the strains on the bars ad and cd, which they severally represent.

The distance dk is the vertical element of the strain dg on the bar $ad = 2$, and dl is the vertical element of the strain on $dc = 6$.

Forces acting in different planes. When three forces whose directions are not in the same plane meet in a point, their resultant is the diagonal of a parallelopiped of which three consecutive edges joining the opposite extremities of the diagonal represent the forces.

Thus the diagonal ad of the parallelopiped (fig. 16) is the resultant of three forces represented by the sides ab, bc and cd. These are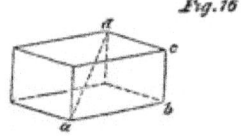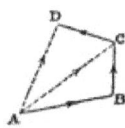

Fig. 16

the equivalent of the three forces shewn by the lines AB, BC and CD, of which the first two only are assumed to be in the same plane, that of the paper. These two may be resolved into one force AC, and the forces AC and CD may be resolved into AD, a force lying in a plane common to both, and the equivalent of the diagonal ad.

In a similar manner the resultant of any number of forces acting in different planes and meeting in a point may be found by taking them in pairs and finding their resultants, again pairing the resultants and so on, until only two forces remain, whose resultant is the required force.

CHAPTER II.

GIRDERS WITH PARALLEL FLANGES.

Load at Centre of Span. Fig. 17 is a representation of a parallel flanged girder of a length l and depth d, loaded at its middle with a weight W. Required the strains on the flanges at the points a and b at the centre of the span.

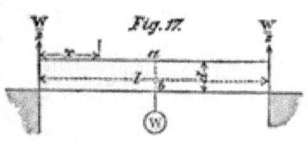

W being half way between the points of support, the upward reactionary force is $\dfrac{W}{2}$ acting at an arm $\dfrac{l}{2}$. The balancing moment of the strains at a and b is Sd, S being the strain, and d, the depth of the girder, being the leverage.

Equilibrium requires that $\dfrac{W}{2} \cdot \dfrac{l}{2} = Sd$, whence $S = \dfrac{Wl}{4d}$.

The strain at a is compression, that at b tension.

The strain on the flanges at any distance x from the nearest abutment (see fig. 17) will be $\dfrac{W}{2} \cdot \dfrac{x}{d}$; this shews that the strain on the flanges at any point is inversely proportional to its distance from the centre of the span.

To find the strain on the flanges at the point of application of the load when that point is not at the centre of the span.

Let x be the distance of the load from one bearing, and $l - x$ the distance from the other bearing, l being the span.

Let W' and W'' be the two reactionary pressures on the bearings, and let

$$W' : W'' :: x : l - x,$$

then
$$W'(l - x) = W'' x;$$

which expressions represent the moments of the pressures on the bearings about the point where the load is applied, where the strain on the flanges is

$$S = \frac{W'(l-x)}{d}, \text{ or } \frac{W''x}{d}.$$

Load evenly distributed, strain at centre. To find the strain on the flanges at the centre of the girder when the load is evenly distributed over it.

Let the point a (fig. 18) be the fulcrum. Let W be the distributed load, l the length, and d the depth of the girder as before.

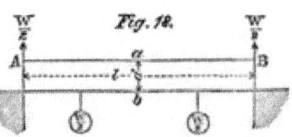

Fig. 18.

Now since the moment of any number of parallel forces about a fulcrum is equal to the sum of the forces acting at their centre of gravity (see p. 3), we may consider the whole load on the part Aa of the girder to be collected a distance half way between A and a, or $\frac{l}{4}$ from either point, its value will of course be $\frac{W}{2}$.

Now the reactionary force of the abutment acts at an arm of $\frac{l}{2}$ about the fulcrum a, its moment therefore is

$$\frac{W}{2} \times \frac{l}{2} = \frac{Wl}{4}.$$

It is balanced by S, the strain on the flange at b acting at a leverage of d, and the force of $\frac{W}{2}$ acting at a leverage of $\frac{l}{4}$;

$$\therefore \frac{Wl}{4} = Sd + \frac{W}{2} \cdot \frac{l}{4}, \text{ whence } S = \frac{Wl}{8d}.$$

This is a very useful formula for finding the strain on the flanges at the centre of a uniformly loaded girder.

Corollary. The strain produced at the centre of the span by an evenly distributed load is one half that produced by the same load collected at the centre of the span.

Strain at any point of the flange. To find the strain on the flanges at any point distant x from the nearest bearing.

Let l be the length of the girder,
" d " " depth " "
" w " " unit of loading,
and wl " " the total load.

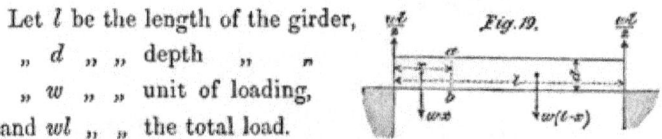

Fig. 19.

Let a (fig. 19) be the point in the flange where the strain is required, and x its distance from the nearest bearing.

Assuming the weight of each of the two portions into which the girder is divided by the point a to be collected at its centre of gravity, the moments about the fulcrum a giving differing signs to the opposing forces will be as follows:

On the left-hand side of the fulcrum,

force	leverage	sign	
wx	$\times \dfrac{x}{2}$	$-$	$= -\dfrac{wx^2}{2}$,
$\dfrac{wl}{2}$	$\times \;\; x$	$+$	$= +\dfrac{wl.x}{2}.$

On the right-hand side of the fulcrum,

force	leverage	sign	
$\dfrac{wl}{2}$	$\times \;\; (l-x)$	$+$	$= +\dfrac{wl(l-x)}{2}$,
$w(l-x)$	$\times \left(\dfrac{l-x}{2}\right)$	$-$	$= -\dfrac{(l-x)^2}{2}.$

Since these moments balance one another,

$$\frac{wlx - wx^2}{2} = \frac{wl(l-x) - w(l-x)^2}{2} = wx\,\frac{l-x}{2}.$$

S being the strain on the flanges at a,

$$Sd = wx\,\frac{l-x}{2}, \text{ and } S = wx\,\frac{l-x}{2d},$$

an expression for the strain on the flanges of a uniformly loaded parallel flanged girder at any distance x from a bearing point.

Example. A girder 12 feet long and 1 foot in depth is loaded with two tons per foot of its length, required the strain on the flanges at a point 3 feet from the bearing.

Here $l = 12$, $d = 1$, $x = 3$, $w = 2$.

Giving these values to l, d, and x in the expression $wx \dfrac{l-x}{2d}$, we get

$$2 \times 3 \left\{ \dfrac{12-3}{2 \times 1} \right\} = 27, \text{ the strain required.}$$

If the strain on the flanges be calculated at a few points in the girder, and a perpendicular proportional to the strain thereat erected from each point, a curved line drawn through the extremities of these lines (see fig. 20) will afford a ready means of arriving at the strain on intermediate points by measuring the distance of the flange from the curve in a direction perpendicular to the former.

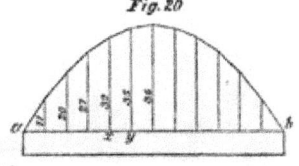

Fig. 20

When a girder is evenly loaded all over, this curve is a parabola of which the apex is situated at the extremity of the perpendicular raised from the centre of the girder, which is the axis of the curve.

In the parabola if from points in any straight line ab (fig. 20) drawn perpendicular to the axis cutting the curve in a and b ordinates be drawn to the curve, their lengths vary as the areas of the rectangles included by the parts into which they severally divide the line ab.

Thus the length of the perpendicular at point x in line ab (fig. 20), is in the same proportion to that at y as $ax \times bx$ is to $ay \times by$.

Now in the equation $S = wx \dfrac{l-x}{2d}$, $\dfrac{w}{2d}$ is a constant quantity; therefore the value of S varies as $x(l-x)$, or the strain on the flanges at any point varies as the rectangle included by the parts into which the length l is divided by the point x, and therefore if the strain on the flange at several points in the girder be expressed by ordinates, the curve which passes through their extremities will be a parabola.

Knowing the strain on the flanges of a girder at one point, we can easily find the strain at any other by the following proportion:

$$S' : S :: (l-y)y : (l-x)x,$$

in which S is the known strain at a point distant x from one extremity of the girder, and S' the required strain at a distance y from the extremity.

Irregularly loaded Girders. To find the strain at any point a in the flanges of a girder at a distance x from one of the bearings when the load is not evenly distributed.

First find the load on the bearing (see p. 33) and then the centre of gravity of the portion x, which we will assume to be at a distance y from the point a.

Let W be the load on the bearing,

and W' „   „   „   portion x.

Then the moment about the point a will be

$$Wx - W'y = Sd;$$

$$\therefore S = \frac{Wx - W'y}{d} = \text{the strain on the flanges at } a.$$

Strain on the Web. Shearing Force. We have now to consider the strain to which the web of a girder is subjected: this will depend upon the amount of *vertical or shearing force*, as it has been called, to which it is subjected.

The vertical or shearing force at any vertical as ab (fig. 19) in a girder is the total vertical downward effect of all external forces on the one side of the plane, which of course equals the total vertical upward effect on the other side of it.

Required the shearing force at the plane ab (fig. 19).

The upward vertical force on the left-hand side of ab is

$$\frac{wl}{2} - wx.$$

The downward vertical force on the right-hand side of ab is

$$w(l-x) - \frac{wl}{2}.$$

GIRDERS WITH PARALLEL FLANGES. 17

The shearing force or F

$$= \frac{wl}{2} - wx = w(l-x) - \frac{wl}{2} = w\frac{l-2x}{2}.$$

From this it appears that in a uniformly loaded girder the shearing force on the web at any point equals the total load between that and the centre of the girder. At the centre of the span all shearing stress disappears.

To find the value of F at any plane ab *distant* x *from one bearing when the girder is not uniformly loaded.*

Find the load on the bearing (which may conveniently be done by finding the centre of gravity of the load) and the weight of the portion between it and the plane ab; the difference between these loads is the shearing force.

Cantilevers. When a beam or girder supports a load which does not lie between the points of support, it is called a cantilever.

Load at the extremity. Fig. 21 is a representation of a cantilever in which the load W is situated at one extremity a of the girder, the other extremity c and an intermediate point b being the bearings.

Here the upward pressure on the bearing at $c = W\frac{ab}{bc}$, and the load on the bearing at $b = W + W\frac{ab}{bc}$.

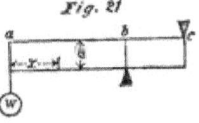

Fig. 21

The strain on the top flange at $b = W\frac{ab}{d}$. The strain on the flange at any distance x from the extremity of the cantilever

$$= W\frac{x}{d}.$$

The vertical or shearing force at any point in the cantilever between a and $b = W$, between b and $c = W\frac{ab}{bc}$.

Distributed Load. When the load is evenly distributed, the strain at b (fig. 22) is equal to that which would result if the total load were concentrated half-way between a and b.

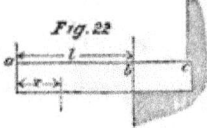

Fig. 22

Y.

If w = unit of loading, l the length of the cantilever, d the depth, the strain on the flange at b or S

$$= \frac{wl \times \frac{l}{2}}{d} = \frac{wl^2}{2d}.$$

The strain on the flange at any distance x from the extremity a is

$$\frac{wx \times \frac{x}{2}}{d} = \frac{wx^2}{2d}.$$

The shearing force at $x = wx$.

In the above two cases the strain on the top flange is tension, that on the bottom compression.

The preceding investigations are chiefly applicable to the cases of girders or beams supporting evenly distributed dead weights, such as the walls of houses or floors of warehouses.

Wooden Beams. In applying the formulæ to wooden beams in which the whole of the section of the beam is considered to resist the bending moment, the upper half by compression, the lower by tension, we must find a new value for d.

Neutral Axis. If, as is usual, we assume that the strain on the fibres increases in direct proportion to their distance from the neutral axis or neutral plane, which is the part of the beam where the longitudinal strains of tension and compression merge into one another, and that this is at the centre of the beam; the forces in compression and tension on each side of it may be represented by shaded wedges (fig. 23) the centres of gravity of which are at the distance $\frac{1}{3}d$ from the neutral axis, and therefore at a distance $\frac{2}{3}d$ apart, if d be the whole depth of the beam.

Fig. 23

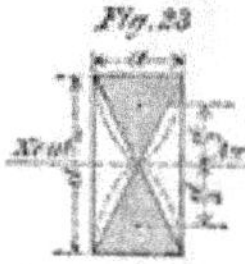

The moment of resistance to bending will therefore be the sum of the strains on the fibres on one side of the neutral axis $\times \frac{2}{3}$ the depth of the beam.

GIRDERS WITH PARALLEL FLANGES.

Thus, let a = the breadth of the beam in inches,
" d = " depth   "   "   "
" $\dfrac{ad}{2}$ = " sectional area of half the beam in inches,
" s = " strain per sq. in. on the outermost fibre.

Then
$$S = \frac{ad}{2} \cdot \frac{s}{2} = \frac{ads}{4},$$

and substituting $\dfrac{2}{3}d$ for d in the expression $\dfrac{Wl}{8d}$ (see p. 13),

$$S = \frac{Wl}{8\frac{2}{3}d} = \frac{3Wl}{16d},$$

therefore
$$\frac{ads}{4} = \frac{3Wl}{16d},$$

whence
$$s = \frac{3Wl}{4ad^2},$$

and
$$W = \frac{4ad^2 s}{3l}$$

= the load which the beam will carry.

Actual experiment proves that timber beams will carry a much greater load than theory would lead us to suppose. The reasons for this are still involved in mystery. One of them, there can be little doubt, is the fact that the fibres in the neighbourhood of the neutral axis are strained much more than is generally supposed, and therefore that, in fig. 23, the shaded part representing the strain on the fibres would more correctly have been bounded by the curved dotted lines, than by the straight lines as in the figure.

The Web in Iron Girders. In iron girders the resistance to bending offered by the web is not usually taken into account, and d is taken to be the distance between the respective centres of gravity of the top and bottom flanges, as shewn in fig. 24. The web however does assist the flanges in resisting transverse strain, and in some cases it may be necessary to determine the amount of this assistance.

Fig. 24

When the girder is of wrought iron, a material which contracts under compression and extends under tension in about an equal ratio, the top and bottom flanges being of equal area, the neutral axis is situated half-way between the flanges, and the formula $W = \dfrac{4ad^2s}{3l}$ (see p. 19) will enable us to ascertain the proportion of the load which may be considered to be supported by the web.

Should the material of the girder be one which compresses more easily than it extends, or the reverse, the position of the neutral axis will not be equidistant between the flanges if they are of equal area.

Fig. 25 represents a portion of beam composed of a material which offers 3 times as much resistance to compression as to extension. Let the line ab represent an imaginary vertical section through the beam. Let de be the position assumed by the line ab when the beam is loaded, the distance ad representing the amount of compression in the upper layer of the fibres, and eb the amount of extension in the lower layer. The point c where these lines intersect lies in the neutral axis. Since the triangles adc and bec are similar to each other,

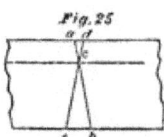

Fig. 25

$$ac : bc :: ad : eb :: 1 : 3,$$

therefore the distance of the neutral axis from the upper or under surfaces of a rectangular beam is inversely proportional to the resistance per unit of section of the fibres at those surfaces to the strain which they experience.

Examples. No. 1. What amount of evenly distributed load will a beam 12 inches deep and 6 inches wide laid across an opening 10 feet wide support, the strain on the outermost fibres to be at the rate of 1 ton per square inch?

Here $a = 6$ in., $d = 12$ in., $s = 1$ ton, $l = 120$ in.,

therefore $W = \dfrac{4 \times 6 \times 144 \times 1}{3 \times 120} = 9\cdot6$ tons.

No. 2. A wrought-iron girder is 1 foot in depth and 12 feet span. The web at the centre of the span is $\tfrac{1}{2}''$ thick; what distributed load is the web capable of sustaining, taking it as 10″

deep over all, and the strain on the outermost fibre as 5 tons per square inch?

Here $a = .5$ in., $d = 10$ in., $s = 5$ tons, $l = 144$ in.

$$W = \frac{4 \times .5 \times 100 \times 5}{3 \times 144} = 2.314 \text{ tons.}$$

Beams and Girders of Irregular Section. To find the load which a beam or girder of irregular section will sustain, the location of the neutral axis must be determined, which will be such that the total moments about the neutral axis of all the forces on the one side of it balance the moments of all the opposing forces on the other side; d the working depth of the beam or girder will be the distance between the centres of gravity of the two sums of opposing forces.

When the section of the beam or girder cannot be equally and symmetrically divided by a line parallel to the line of action of

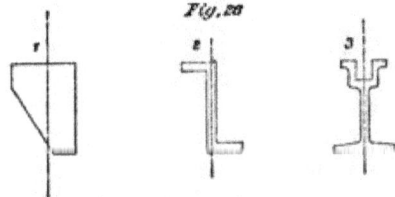

Fig. 26

the load, generally a vertical line, allowance must be made for the tendency to buckle unless the beam or girder be prevented from so doing by stays.

Sections 1 and 2, fig. 26, are examples of this want of symmetry. No. 3 is symmetrical.

Practical method of determining the strain on an ordinary wrought-iron girder.

Ordinary wrought-iron Girders. In the commonest type of wrought-iron girder bridge, the roadway is carried by cross girders which are supported by main girders. The points of attachment of the cross girders, whether they are suspended from the bottom or rest upon the top flange, are stiffened in the main girder by upright gusset plates or T irons. The load is considered to be concentrated at these points, and the strain on the flange between two adjacent gussets to be uniform.

Diagram 1, Plate I., is a diagram for a girder in which the cross girders attached to the bottom flange are equally loaded, and placed at equal distances apart. The web of the girder is assumed to be so thin as to be incapable of resisting compression, and therefore to do duty in tension only.

Let the load at each gusset be 8 tons.

Since the girder is uniformly loaded, there will be an equal load of 44 tons on each pier, consequently of the load of 8 at the centre of the span 4 will be borne by the right-hand, and 4 by the left-hand abutment. The load 4 is supported by the web represented by the diagonal line, and is conveyed up to the point where the adjoining gusset meets the top flange. Here it puts a horizontal compressive strain on the top flange, and a vertical strain of 4 on the gusset. This force of 4 being added, the load of 8 makes a total load of 12 to be carried by the web in the 2nd bay from the centre. In this manner the load on the web in each bay increases towards the pier.

Strains on the Flanges. The strains on the flanges in each bay are found by commencing with the bay next the abutment and working towards the centre of the span.

For the sake of convenience, the depth of the girder has been taken equal to the width of a bay and therefore the horizontal strain on the top flange on the last bay is 44, which is also the vertical load on the pier, for the point a is kept in equilibrium by three forces represented by the sides of a right-angled triangle, in which the two sides enclosing the right angle are equal to one another.

In like manner the point b in the bottom flange is kept in equilibrium by three forces, of which the vertical is 44, consequently the horizontal strain on the bottom flange of the second bay from the abutment is 44, being equal to the vertical.

These horizontal strains on the flanges are carried through till they meet with reacting strains coming from the other end of the girder.

The vertical load of 36 carried by the diagonal of the second bay adds 36 to the strain on the flanges, making it 80 + in the top flange, and 80 − in the bottom.

Adding in the horizontal effect of each diagonal as we proceed towards the centre we thus get the strain on the flanges in each bay, and a maximum strain of 144 tons at the centre.

Diagram 2, Plate I., shews the strains on the flanges, the web being assumed to act *only* in compression, while Diagram 3, Plate I., gives the strains which result when the web resists equally extension and compression.

It may be observed that the *sum* of the strains on the top and bottom flange of any bay is the same for each of the three kinds of web, while in the last case the strains on the top and bottom flanges are equal, being the mean of the strains on the two flanges by either of the other methods.

To prove the truth of the third diagram as in Diagrams 1 and 2, Plate I., take the load per bay as 4 instead of 8; find the strains on the flanges, which will be just half those given in the figure, and consider Diagram 1 to be placed upon Diagram 2, so that the flanges and verticals of each coincide, and the strains of each to be added together: the result will be Diagram 3, Plate I.

In practice, find the strains by the method of either Diagram 1, or Diagram 2, and take the mean as the correct strain for the flanges, the web being strained equally in tension and compression.

Since As depth of girder : breadth of bay
:: vertical load on a diagonal : horizontal strain caused thereby,

∴ horizontal strain = vertical load × $\dfrac{\text{breadth of bay}}{\text{depth of girder}}$, or $\dfrac{b}{d}$,

if $b =$ breadth of bay,
and $d =$ depth of girder.

In the example we have just considered $\dfrac{b}{d} = 1$.

The strains on the flanges in Diagrams 1, 2, and 3, multiplied by the fraction $\dfrac{b}{d}$, will give the correct strains for any similarly loaded girders in which the proportion of the depth to the breadth of a bay is as d to b.

GIRDERS WITH PARALLEL FLANGES.

Strains on the Web. The web in plate girders should be considered to act as in Diagram 3, in this case it is made to do double duty; the same fibres which are resisting extension in one direction, resist a force of compression in a direction at right angles to the other, without impairing the efficiency of the plate in offering resistance to either force.

On the system of Diagram No. 3 the webs need only be one-half the thickness of those required for Diagrams Nos. 1 and 2.

It is a further advantage attending the system of Diagram 3 over Diagrams 1 and 2, that it allows the top and bottom flanges to be made exactly alike without waste of metal.

Diagram No. 4, Plate I., is similar to Diagram No. 1, but the letter w is substituted for the load of 8 on each bay. It shews a more expeditious way of finding the strains on the flanges, saving figures.

Adding the top and bottom flanges together in any bay, and dividing by 2, we get an expression for the strain on the system of Diagram No. 3.

Thus the strains on the third bay from the abutment on this system would be $\dfrac{13\frac{1}{2}\ W + 10\ W}{2} = 11\frac{3}{4}\ W = 94$, if $w = 8$.

Result checked by method described p. 14. The strains on the flanges in any bay may also be found by the method indicated at the commencement of this Chapter, for finding the strain on the flanges at any point of a uniformly loaded girder. We cannot make use of the formula $S = wx\,\dfrac{l-x}{2d}$, as our load is not evenly distributed, but concentrated at intervals.

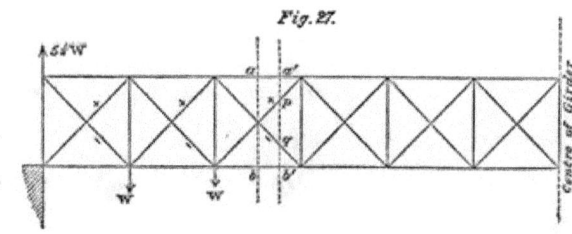

Fig. 27.

Taking a girder of which Diagram No. 3, Plate I., is a diagram, required the strain on the flanges of the third bay from the abutment. Draw the vertical line ab (fig. 27) through the centre of the bay, and therefore passing through the point of intersection of the diagonals.

The fulcrum being supposed to be either the point a or b, we have an upward force representing the reaction of the pier $= 5\frac{1}{2} W$, which acts at a leverage of $2\frac{1}{2}$, if the length of a bay be taken as 1. In opposition to this we have two forces, W acting at a leverage of $\frac{1}{2}$, and again W at a leverage of $1\frac{1}{2}$.

The difference between the moments of these forces is the actual bending moment about point a or b.

Since the length of each bay is 1, and the depth 1,

The moment of the reaction of the abutment is......... $5\frac{1}{2} W \times 2\frac{1}{2} = 13\frac{3}{4} W +$

The moment of the load between the fulcrum and the abutment is................ $\left. \begin{array}{c} W \times \frac{1}{2} \\ W \times 1\frac{1}{2} \end{array} \right\} = 2W$

and ....

Bending moment or $S = 11\frac{3}{4} W$

If $W = 8$, $S = 94$.

It is proper to consider the flanges throughout the whole bay as subject to a strain of $11\frac{3}{4} W$, though if the plane ab in which the fulcrum lies were taken at a point in the bay nearer the centre of the girder, the value of S would be greater, while on the other hand it would be less if it cut the bay at a point nearer to the abutment than the centre of the bay. In such case though the strain on the *flange* would be the same, since the line ab does not pass through the point of intersection of the diagonals, *their* effects would have to be reckoned in determining the total bending moment. The moments of the diagonals depend upon the vertical distance apart of the points where they are intersected by the imaginary plane.

Example. In the girder just given, required the total bending moment on S at a distance of $2\frac{3}{4}$ bays from the abutment, being at the section $a'b'$, fig. 27.

Proceeding as in the previous example,

Moment of reaction of abutment $= 5\frac{1}{2} \; W \times 2\frac{3}{4} = 15\frac{1}{8} \; W +$

Moment of loads at each bay $= \begin{Bmatrix} W \times 1\frac{3}{4} \\ W \times \dfrac{3}{4} \end{Bmatrix} = 2\frac{1}{2} \; W -$

$$S = 12\frac{5}{8} \; W$$

when width of each bay and depth of girder, or $d, = 1$.

The value of S for the flanges of bay No. 3 we have seen to be $11\frac{3}{4} W$, therefore

$$12\frac{5}{8} W - 11\frac{3}{4} W = \tfrac{7}{8} W,$$

which must be the bending moment of the diagonals acting at a leverage of d.

Now the line $a'b'$ intersects the diagonals in the points p and q and the distance apart of these points is half the depth of the girder, or $\dfrac{d}{2}$.

The horizontal effect of each diagonal in bay No. 3 is equal to its vertical effect, since the diagonals are inclined at an angle of 45°.

By referring to Diagram No. 4, Plate I., it will be seen the total vertical or shearing force on bay No. 3 is $3\frac{1}{2} W$. The vertical force on each diagonal when they are crossed is therefore $1\frac{3}{4} W$, which is also the amount of horizontal force exerted by each, the diagonal in which the point p lies giving a compressive horizontal force and the other a tensive force.

Now $1\frac{3}{4} W$ acting at a leverage of $\dfrac{d}{2}$

$= \tfrac{7}{8} W$,, ,, ,, d;

therefore considering all the horizontal forces as acting at a leverage of d,

$$11\frac{3}{4} W + \tfrac{7}{8} W = 12\frac{5}{8} W,$$

the value of S at a distance of $2\frac{3}{4}$ bays from the abutment.

In finding the value of S at a point distant $2\frac{1}{4}$ bays from the abutment, the horizontal effect of the diagonal has to be *deducted* from the flange, as the strain on the diagonal is of a different sign to that on the adjacent flange.

General Formula for the Strain on any Bay. By means of the foregoing diagrams formulæ may be constructed for finding the strains on the flanges of any bay of a girder.

When loaded as per Diagram No. 1, Plate I., the web being in tension only,

Let N = whole number of bays in the girder,
$\quad n$ = number of any bay reckoned from the abutment,
$\quad b$ = breadth of a bay,
$\quad W$ = load per bay,
$\quad d$ = depth of girder,
$\quad S$ = strain on flange of any bay n,

$$S = n\frac{N-n}{2}W\cdot\frac{b}{d}, \text{ for top flange} \dots\dots\dots\dots\dots (1),$$

$$S = (n-1)\frac{N-(n-1)}{2}W\cdot\frac{b}{d}, \text{ for bottom flange} \dots\dots (2).$$

When the web is under compression only as in Diagram No. 2, Plate I., equation No. 1 will give the strain on the bottom flange, and equation No. 2 on the top flange.

When the web plate acts equally in compression and extension, the mean of equations 1 and 2 will give the strain on either flange; wherefore

$$S = \frac{n\frac{N-n}{2}W + (n-1)\frac{N-(n-1)}{2}W}{2}\cdot\frac{b}{d},$$

or $\quad S = \dfrac{Wb}{2\,d}\left\{n(N-n+1) - \dfrac{N+1}{2}\right\}$ ....(3).

Shearing strain on each Bay. The method described on p. 16 would enable us to determine the shearing force on any bay of a girder loaded at intervals, but as for each alteration in the position of the moving load the proportion of load borne by each abutment has to be calculated afresh, a more expeditious method of ascertaining the maximum shearing strain on any bay of the girder is to construct a series of diagrams as in Plate II., in which the shearing effects produced by the different positions of the load are shewn in separate diagrams.

The diagrams 1 to 14, Plate II., represent a girder of 12 bays, having a dead load of 2 tons, and a live load of 6 tons per bay.

Diagram No. 1 shews the shearing strains produced by the dead load which are constant, the diagonal line across each bay serves to shew by which abutment the load on that bay is carried, the upper end of the line lying towards the particular abutment.

The diagonal lines may be regarded as *tension* rods and the figures written against them as the *vertical* elements of the strains to which they are subjected.

Diagram No. 2 shews the strains caused by loading the 1st bay, counting from the left hand, with 6 tons.

Diagram No. 3, the strains caused by the loading of the 2nd bay, and so on.

The strain on bay No. 2, Diagram No. 3, is 2·25 only; this is the *difference* between 2·5, the proportion of the load of 3 resting at the junction of bays 2 and 3, which goes to the *left* hand abutment; and ·25, the proportion of the load of 3 situated between bays 1 and 2, which goes to the *right* hand abutment. For this bay diagonals crossing one another having the numbers 2·5 and ·25 written against them respectively would represent the shearing forces due to each of the loads of 3 in turn. In all cases where diagonals representing direction of shearing stress cross, the only vertical force upon the bay is the excess of the preponderating diagonal over the other, and the only diagonal to be shewn is the one representing the intenser force of the two.

If the vertical element of two diagonals crossing one another is the same, the shearing strain on the bay is nil: such is the case in the central bay of a symmetrically loaded girder with an odd number of bays.

The propriety of *subtracting* opposing diagonal effects may further be seen if we remember that shearing force produces strain

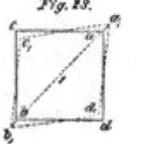
Fig. 13.

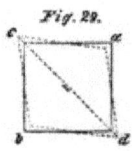
Fig. 29.

GIRDERS WITH PARALLEL FLANGES.

on the diagonal, consequently distortion of the normal form of the bay. When the diagonal ab (fig. 28) is subjected to tension it stretches; in consequence of this the originally square form of the bay $abcd$ (fig. 28) becomes a rhombus $a_1 b_1 c_1 d_1$, the points a and b being forced further apart and the points c and d brought closer together.

When the diagonal cd (fig. 29) is in tension, a reverse action takes place, so that were the respective hypothetical strains on the diagonals ab and cd equal, the rectangular form of the bay would be preserved, in which case there can be no actual strain on the diagonals. If the hypothetical strain on the one exceed that on the other, the actual strain will be the difference of the two.

Diagram No. 14 exhibits the greatest shearing force for every bay of the girder. The 4 centre bays have two sets of figures and diagonals crossing one another: this signifies that under one condition of loading the direction of the diagonal differs from that under another possible condition of loading. The figures written against each diagonal shew the vertical component of the greatest strain to which the diagonal can be exposed.

In Diagram 14, the vertical strain of 44 in bay No. 12 is obtained by taking the sum of the strains on bay No. 12 in all the diagrams 1 to 13 inclusive. In bay No. 11 the strain of 36·25 is the sum of the strains on bay No. 2, in Diagrams 1 to 12 inclusive, that is to say the live load is supposed to be removed from bay No. 12, the girder being loaded throughout with that exception. When the girder is loaded all over the strain on bay No. 11 is 36, since ·25 the strain on the opposing diagonal of bay No. 11, Diagram 13, has to be *deducted* from 36·25.

Similarly in bay No. 10 the shearing strain is 29, being the sum of the strains in Diagrams 1 to 11 inclusive.

In bay No. 7 the strain is 5·25, bays Nos. 8 to 12 being loaded. 5·25 is the sum of the strains on bay No. 7 in Diagrams 9 to 13 inclusive, − 1 *the strain on bay No. 7 in Diagram No.* 1. Since the dead load is always constant, this neutralizing strain must not be neglected.

The shearing strain on bay No. 7 when bays 1 to 7 are loaded will be 10·25, being the sum of the strains in Diagrams 1 to 8 inclusive, the direction of the diagonal in this case being reversed.

The shearing strain on bay No. 5 caused by loading bays No. 1 to No. 4 is 1 only. It is the sum of strains on bay No. 5 in Diagrams 2 to 5 less 3, the opposing strain on bay No. 5 in Diagram 1.

There is no shearing force to left abutment on bay No. 9, for if we add together the strains on bays No. 9 in Diagrams 11 to 13 we get a shearing force to left hand of 2·25, but the shearing force to the right-hand abutment on bay No. 9 caused by the dead load is 5 as given in Diagram 1. Subtracting these forces from one another we have a remaining shearing force of 2·75 to the right hand.

Hence it is evident that if Diagram 14 represents a lattice girder with vertical struts and diagonal tension bars, bars crossing one another should be inserted in the 4 central bays. For a plate girder the web in each of the 4 centre bays must be made strong enough to sustain the greater of the two shearing strains. Since the web does duty equally in compression as in tension, the vertical shearing force will be halved for each diagonal; thus in bay No. 1 the web will have to sustain a vertical force of 22 by tension and a like amount by compression.

The bays in this example being squares, the actual strains on the diagonals will be the vertical load upon them $\times \sqrt{2}$.

Formula for Shearing Force. *Formula for the maximum shearing force upon any bay.*

If we examine Diagram No. 4, Plate I., we shall see that the shearing force on the bay next the pier may be expressed by the formula

$$F = W \frac{N-1}{2},$$

W being the load per bay and N the total number of bays in the girder.

In the 2nd bay from the abutment $F = W \dfrac{N-3}{2}$,

,, ,, 3rd ,, ,, ,, ,, $F = W \dfrac{N-5}{2}$,

,, ,, 4th ,, ,, ,, ,, $F = W \dfrac{N-7}{2}$,

and so on. Up to the 6th bay the value of F is positive, but if after reaching the centre of the girder we continue to count the number of the bays from the same abutment, we shall get F with a negative sign; this indicates the change in the direction of the diagonal. The value of F will however be correct, as, for instance, it may be written thus for the bay next the abutment,

either $\quad F = W \dfrac{N-1}{2}$ for bay No. 1 $= W \dfrac{12-1}{2} = 5\frac{1}{2} W$,

or $\quad F = W \dfrac{N-23}{2}$ for bay No. 12 $= W \dfrac{12-23}{2} = -5\frac{1}{2} W$,

according as we count from the right or left-hand abutment. A negative value to F shews that the shearing force is carried by the abutment opposite to that from which the counting has commenced.

Putting n for the number of the bay, the formula for the vertical or shearing force on any bay of a symmetrically loaded girder becomes
$$F = W \dfrac{N-(2n-1)}{2}.$$

By deducting the values of F as given in Diagram 1, from its values in Diagram 14, Plate II., it will be seen that the greatest shearing force caused by a *live load* of W' per bay is—

On bay No. 1, $W' \left(\dfrac{N-1}{2} \right)$,

„ „ „ 2, $W' \left(\dfrac{N-3}{2} + \dfrac{1}{2N} \right)$,

„ „ „ 3, $W' \left(\dfrac{N-5}{2} + \dfrac{4}{2N} \right)$,

„ „ „ 4, $W' \left(\dfrac{N-7}{2} + \dfrac{9}{2N} \right)$,

and so on.

For the expressions $\dfrac{N-1}{2}$, $\dfrac{N-3}{2}$, $\dfrac{N-5}{2}$, &c. substitute the general formula $\dfrac{N-(2n-1)}{2}$, n being the number of the bay, and express $\dfrac{1}{2N}$, $\dfrac{4}{2N}$, $\dfrac{9}{2N}$, &c. by $\dfrac{(n-1)^2}{2N}$.

Thus the general formula for the shearing stress on any bay n due to the live load becomes

$$F = W'\left\{\dfrac{N-(2n-1)}{2} + \dfrac{(n-1)^2}{2N}\right\}.$$

Therefore the maximum shearing force on any bay n caused by the live and dead loads combined will be expressed by the formula

$$F = W'\left\{\dfrac{N-(2n-1)}{2} + \dfrac{(n-1)^2}{2N}\right\} + W\dfrac{N-(2n-1)}{2},$$

or $\qquad F = \dfrac{W+W'}{2}\{N-(2n-1)\} + W'\dfrac{(n-1)^2}{2N};$

when the number of the bays in the girder is *even*. When the number of bays is *uneven*, the formula is true when n is not greater than $\dfrac{N}{2}$; but when this is the case, a quantity represented by the expression $\dfrac{W'}{2N}(N-2n-1)$ must be added to the right-hand side of the above equation to obtain the true value of F. Therefore when the bays are uneven, the value of F in any bay whose number is more than half the total number of bays in the girder may be expressed by the formula

$$F = \dfrac{W+W'}{2}\{N-(2n-1)\} + \dfrac{W'}{2N}\{(n-1)^2 + N + 2n - 1\}.$$

In the foregoing formulas,

$\qquad F\ =$ maximum shearing force on the bay n,
$\qquad n\ =$ the number of the bay counting from one abutment,
$\qquad N\ =$ the total number of bays in the girder,
$\qquad W' =$ the live load per bay,
$\qquad W\ =\ $ „ dead „ „ „

GIRDERS WITH PARALLEL FLANGES. 33

N.B. *Negative* values of F obtained from the above formulas should be neglected. For the bays near the centre of the girder F have *two positive* values, varying accordingly as the number (n) of the bay is counted from one or the other abutment. Of these the greater will be that obtained by counting the number n of the bay from the *nearer* abutment. This value of F will be the only one required for a plate-girder. For a lattice-girder both values of F must be found, and diagonals crossing one another introduced, as shewn in Diagram 14, Plate II.

Irregular loading, First case, Bays equal. In Diagram 5, Plate I. are given the strains on the flanges of one girder of a skew bridge, the load on the left-hand end of the girder being less than that on the right. For simplicity's sake, the web is assumed as acting in tension only.

There are two convenient methods of finding the load borne by each abutment, the first by dealing with each load singly, and adding the separate effects of all the loads as follows:

		Proportion to left-hand abutment	Proportion to right-hand abutment
Of load	1	$\frac{11}{12}$	$\frac{1}{12}$
,, ,,	3	$3 \times \frac{10}{12} = \frac{30}{12}$	$3 \times \frac{2}{12} = \frac{6}{12}$
,, ,,	5	$5 \times \frac{9}{12} = \frac{45}{12}$	$5 \times \frac{3}{12} = \frac{15}{12}$
,, ,,	7	$7 \times \frac{8}{12} = \frac{56}{12}$	$7 \times \frac{4}{12} = \frac{28}{12}$
,, ,,	8	$8 \times \frac{7}{12} = \frac{56}{12}$	$8 \times \frac{5}{12} = \frac{40}{12}$
,, ,,	8	$8 \times \frac{6}{12} = \frac{48}{12}$	$8 \times \frac{6}{12} = \frac{48}{12}$
,, ,,	8	$8 \times \frac{5}{12} = \frac{40}{12}$	$8 \times \frac{7}{12} = \frac{56}{12}$
,, ,,	8	$8 \times \frac{4}{12} = \frac{32}{12}$	$8 \times \frac{8}{12} = \frac{64}{12}$

GIRDERS WITH PARALLEL FLANGES.

	Proportion to left-hand abutment	Proportion to right-hand abutment
Of load 8	$8 \times \dfrac{3}{12} = \dfrac{24}{12}$	$8 \times \dfrac{9}{12} = \dfrac{72}{12}$
,, ,, 8	$8 \times \dfrac{2}{12} = \dfrac{16}{12}$	$8 \times \dfrac{10}{12} = \dfrac{80}{12}$
,, ,, 8	$8 \times \dfrac{1}{12} = \dfrac{8}{12}$	$8 \times \dfrac{11}{12} = \dfrac{88}{12}$
Totals	$\dfrac{366}{12} = 30\cdot 5$	$\dfrac{498}{12} = 41\cdot 5.$

Since the total load on the left-hand abutment is 30·5
and ,, ,, ,, ,, right-hand ,. ,, 41·5,
we must so divide the loads that the sums of those which go to the left shall be 30·5, and the sum of those which go to the right 41·5. To do this, assume that 6·5 of the load of 8 at the centre of the girder goes to the left, and 1·5 of it to the right-hand abutment,
then $1 + 3 + 5 + 7 + 8 + 6\cdot 5 = 30\cdot 5$ load on left-hand abutment
and $8 + 8 + 8 + 8 + 8 + 1\cdot 5 = 41\cdot 5$,, ,, right-hand ,.

The moments of these loads balance one another.

Moments of Loads

On left-hand abutment

load	leverage	moment
1	× 1 =	1
3	× 2 =	6
5	× 3 =	15
7	× 4 =	28
8	× 5 =	40
6·5	× 6 =	39
Total moment		129

On right-hand abutment

load	leverage	moment
8	× 1 =	8
8	× 2 =	16
8	× 3 =	24
8	× 4 =	32
8	× 5 =	40
1·5	× 6 =	9
Total moment		129

The second method is to find the moments of all the loads about one abutment which are equal to the moment of the pressure on the *other* abutment about the first abutment. Thus the moments about the left-hand abutment are

$$
\begin{array}{r}
1 \times 1 = 1 \\
3 \times 2 = 6 \\
5 \times 3 = 15 \\
7 \times 4 = 28 \\
8 \times 5 = 40 \\
8 \times 6 = 48 \\
8 \times 7 = 56 \\
8 \times 8 = 64 \\
8 \times 9 = 72 \\
8 \times 10 = 80 \\
8 \times 11 = 88 \\ \hline
498
\end{array}
$$

Let P be the pressure on the right-hand abutment, P acts at an arm of 12,

$$\therefore P \times 12 = 498,$$
$$P = \frac{498}{12} = 41\cdot 5,$$

which agrees with result obtained by the first process.

Second Case, Bays of an unequal size, Load symmetrical. Diagram 6, Plate I. is the diagram for a cross girder 25 feet long and 2 feet deep. The load being symmetrically disposed on each side of the centre line, there will be no shearing strain on the centre bay under full load, but an equal load of 16 on each abutment. Assuming the web to act only in tension, the strains on the flanges would be those given on the left-hand side of the centre line of the diagram. On the right-hand side are given the strains which arise when the web acts equally in tension and compression.

The strain of 36 on the flange of the 4'. 6" bay to the left is obtained by multiplying 16, the vertical load carried by the diagonal, by 4'. 6" the length of the bay, and dividing by 2 the depth of the bay, 4·6 being the leverage at which the load acts, and 2 feet the leverage of the counterbalancing strain on the flange. The method by the parallelogram of forces also shews us that the forces which keep the upper left-hand corner of the girder in equilibrium are proportional to the sides of a triangle composed of the depth of the girder 2', the length of the bay

4′. 6″ and the diagonal of the bay, since the depth of the girder 2 represents the vertical force 16.

∴ As $2' : 16 :: 4'. 6'' : 36$, the horizontal strain on the flange.

It may be shewn in a similar manner that $8 \times \frac{5}{2} = 20$ is the additional strain on the flange caused by the load of 8 at the junction between the second and third bays, and $36 + 20 = 56$ is the strain on the top flanges of second and third bays.

On page 23 is given the following equation for the strain on the flanges caused by a vertical load on any bay. Horizontal strain = vertical load $\times \frac{b}{d}$.

When the bays of a girder are unequal in length, the fraction $\frac{b}{d}$ will vary for each bay. This is the only point of difference between the case we have just considered and that of the uniformly loaded girder treated of on page 22.

Third Case, Bays and Loading irregular. In Diagram 7, Plate I. the bays and loads are both irregular.

Proceeding as in the case of Diagram 5, Plate I.

	Proportion to left-hand abutment	Proportion to right-hand abutment
Of load 8	$8 \times \frac{23}{25} = \frac{184}{25}$	$8 \times \frac{2}{25} = \frac{16}{25}$
,, ,, 6	$6 \times \frac{19}{25} = \frac{114}{25}$	$6 \times \frac{6}{25} = \frac{36}{25}$
,, ,, 12	$12 \times \frac{15}{25} = \frac{180}{25}$	$12 \times \frac{10}{25} = \frac{120}{25}$
,, ,, 9	$9 \times \frac{9}{25} = \frac{81}{25}$	$9 \times \frac{16}{25} = \frac{144}{25}$
,, ,, 4	$4 \times \frac{4}{25} = \frac{16}{25}$	$4 \times \frac{21}{25} = \frac{84}{25}$
Totals	$\frac{575}{25} = 23,$	$\frac{400}{25} = 16.$

To get a load of 23 on the left-hand abutment, and of 16 on the right-hand abutment, we must split the load of 12, taking 9 to the left and 3 to the right-hand side.

The moments of loads are as follows:

On left-hand abutment	On right-hand abutment
load leverage moment	load leverage moment
8 × 2 = 16	4 × 4 = 16
6 × 6 = 36	9 × 9 = 81
9 × 10 = 90	3 × 15 = 45
Total moment 142	Total moment 142

This moment of 142 divided by 2, the depth of the girder which represents the arm or leverage at which the strain on the flange acts, gives 71 as the accumulated or maximum strain on the flanges of the girder.

In order to find the greatest shearing stress on any bay of girder with irregular bays, caused by a moving load, the method illustrated by Plate II. should be adopted.

Trusses and Lattice Girders. If for a solid plate web we substitute open work of bars, whether acting as ties or struts, inclined at an angle with the vertical, we have what is called a *truss* or a *lattice* girder.

Diagram 1, Plate II. may be taken to represent a simple truss in which the diagonals are tension-bars, and the verticals struts, and Diagram 2 a truss in which the diagonals are struts, and the verticals tension-bars. When several sets of diagonal bars crossing one another are used, the girder is styled a *lattice*, and sometimes a *trellis* girder, from its resemblance to trellis-work.

Without Verticals. Diagram 1, Plate III. is a diagram of a truss in which there are no verticals; the diagonals are inclined at an angle of 45° with the vertical. The left-hand half of the diagram gives the strains on the flanges, and the vertical stress on the diagonals caused by resting the load on the top flange. The right-hand portion gives the strains when the load is suspended from the bottom flange.

It will be observed that the strains may be made to agree in the two parts by turning either half upside down and changing the signs of the strains.

The formula

$$S = \frac{W}{2}\frac{b}{d}\left\{n(N-n+1) - \frac{N+1}{2}\right\}$$ (see page 27),

will give us the *mean strain*, i.e. *half the sum of the strains* on the top and bottom flanges of any bay n.

When the load *rests upon the top flange*, the strain on the bottom flange will exceed that on the top flange of the same bay by the amount $\frac{W}{2}$. But when the load *is suspended from the bottom flange*, the strain on the bottom flange will be less than that on the top flange by the amount $\frac{W}{2}$; accordingly the true strain on the flanges will be obtained, as the case may be, by adding $\frac{W}{4}$ to, or subtracting $\frac{W}{4}$ from, the value of S obtained from the formula above.

Plate IV. shews by a series of diagrams the shearing effect of the line load considered at each of its points of application in succession.

Diagram 14 gives the extreme vertical strains. From this diagram it will appear that the diagonals at the centre of the truss have to be made capable of resisting both tension and compression.

The *sums* of the extreme strains on the diagonals of any bay n may be found by the formula

$$F = \frac{W+W'}{2}\{N-(2n-1)\} + W'\frac{(n-1)^2}{2N}$$ (see page 32).

In each bay the vertical strain on the one diagonal exceeds that on the other by $\frac{W+W'}{2}$.

When the load is suspended from the bottom flange the diagonal in tension is subject to the greater strain, but if the load rest on the top flange the reverse is the case; see Diagram 1, Plate III.

Warren Truss. The peculiarity of the Warren truss is that the struts and ties make angles of 60° or thereabouts with the flanges.

GIRDERS WITH PARALLEL FLANGES. 39

In Diagram 2, Plate III. are given the strains on a Warren truss in which the load rests upon the top flange. Diagram 3, Plate III. shews the strains caused by suspending the load from the bottom flange. For the sake of comparison with the diagrams of girders and trusses already given, the depth of these Warren trusses has been taken as $\frac{1}{12}$ of the span, consequently the angle made by the diagonals with the flanges is more than $60°$, being about $63°$.

Referring to Diagram 3, Plate I. it will be seen that the strains on the bottom flange of that diagram agree with those on the top flange of Diagram 2, Plate III. and with those on the bottom flange of Diagram 3, Plate III. These strains may consequently be found by equation 3, page 27.

The strains on the eleven bays in the bottom flange of Diagrams 2 and 4, and in the top flange of Diagrams 3 and 5, Plate III. may be found by the equation

$$S = n \frac{N-(n-1)}{2} W \frac{b}{d}; \text{ see p. 27;}$$

N being 11 in this case.

The strains on the top flange of Diagram 4, and bottom flange of Diagram 5, Plate III. are given by the formula

$$S = \frac{W}{2} \frac{b}{d} \left\{ n(N-n+1) - \frac{N}{2} \right\},$$

and on the bottom flange of Diagram 3, Plate III. by equation 3, p. 27.

In the example Diagram 6, Plate III. the load is supported by both top and bottom flanges. It may be regarded as a combination of the two systems illustrated by Diagrams 3 and 5, Plate III.

Since the load at each point of intersection of diagonals with the flanges is 4 in Diagram 6; we may, in order to find the strains, take 4 as the load per bay in each of the Diagrams 3 and 5, calculate the strains, and consider the one to be superimposed upon the other, adding the strains.

The strains on the top flange of Diagram 6, Plate III. may be found by the equation above given for the top flanges of Diagrams 3 and 5. For the bottom flange the strains may be found by the formula

$$S = \frac{W}{2}\frac{b}{d}\left\{n(N-n+1) - \frac{2N+1}{4}\right\}.$$

It must be remembered that N is the whole number of bays (12 in the case of Diagram 6, Plate III.) and W the load per bay ($=8$ in this case).

Strains on the Diagonals. The strains on the diagonals of a Warren truss may be found by constructing a series of diagrams as in Plate II., or by means of the formula given on page 32.

Diagram 7, Plate III. shews the strains on the diagonals of a Warren truss with a live load of 6, and a dead load of 2 per bay.

CHAPTER III.

HOGBACKED GIRDERS.

We now come to the consideration of girders in which the flanges are not parallel. Girders which have a slightly curved top flange are very common, these have been called Hogbacked girders. The effect of thus diminishing the depth towards the abutments is to increase the strain on the flanges and relieve the web.

Equations 1, 2 and 3, p. 27, will give the strains on the flanges of any bay of a hogbacked girder according as the diagonals do duty in tension only, compression only, or in tension and compression alike. The fraction $\frac{b}{d}$ in those equations will vary for each bay.

Diagrams 1 to 6, Plate V. represent a hogbacked truss with diagonal tie-bars under various conditions of loading. Diagram 7 gives the extreme horizontal strains on the diagonals, and the greatest strains on the verticals; Diagram 1 the extreme horizontal strains on the flanges.

Strains on the Flanges. The horizontal strains on the flanges may be found conveniently by the method described on pp. 24 and 25. Thus at the points a and b, Diagram 1,

$$S = 45 \times \frac{1}{.775} = 58.06;$$

at points c and d,

$$S = 45 \times \frac{2}{.95} - 18 \times \frac{1}{.95} = 75.79;$$

at points e and f,

$$S = 45 \times \frac{3}{1} - \frac{18 \times (2+1)}{1} = 81.$$

Strains on the Diagonals. To find the greatest strain on the diagonals, it will be necessary to calculate the strains on the flanges of the girder under the various conditions of loading.

The extreme strains on the diagonals of bays Nos. 1 and 6 occurs when the girder is fully loaded, as in Diagram 1, in which case the horizontal element of the strain on the diagonal is obviously 58·06, as it forms the medium of communicating the horizontal strain of 58·06 in the top flange to the bottom flange at b.

The greatest strain occurs on bay No. 2, when the bays Nos. 2 to 6 of the girder are fully loaded, as shewn in Diagram 2.

The horizontal strain on the diagonal of this bay is equal to the difference between the horizontal strains on the top flange of bays 1 and 2, or between the strains on the bottom flange of bays 2 and 3.

This is obvious, for to maintain point g or point k in equilibrium, the opposing horizontal as well as vertical forces upon g must be equal; for instance, the horizontal thrust upon g by the top flange in bay 2 being 71·58, it is opposed by 51·61 in bay No. 1, and 71·58 − 51·61, or 19·97 must be the horizontal pull of the diagonal to preserve equilibrium.

Thus, *the horizontal strain on a diagonal is equal to the difference between the horizontal strains on the flanges on each side of the point of its junction therewith.*

Having found the horizontal element of the strain on the diagonal, it is easy to calculate therefrom the vertical element; this determines the strain on the vertical: e.g. take point b, Diagram 1, which is subjected to an upward vertical pull of 29·03 from the diagonal, while a load of 18 tends to pull it vertically downwards, to maintain equilibrium therefore we require a force of 29·03 − 18, or 11·03 compressive strain on the vertical. On the other hand, to keep point d in equilibrium an upward force of 18 − 13·735 or 4·265 is required *in tension* on the vertical.

Diagram 7 contains the maximum strains on diagonals and verticals collated from Diagrams 1 to 6. Diagrams 1 to 6 give the strains on every part of the truss under each condition of loading. The correctness of the figures given may be checked by examining, in turn, each point where the diagonals and verticals meet the flange, and observing whether the conditions of equilibrium are maintained.

For instance, point a Diagram 1 is acted upon by a horizontal force of 75·79 which is balanced by two horizontal forces 58·06 and 17·73, whose sum equals 75·79: thus equilibrium as regards horizontal motion is proved. Again, the vertical forces tending to force point a downwards are 13·265 and 13·735, their sum $= 27$. These are balanced by the forces 15·97 and 11·03, sum $= 27$, tending to force point a upwards: thus equilibrium as regards vertical motion is proved, and the correctness of the figures ascertained.

A further check upon the figures may be obtained by adding together the total vertical effects of flange and diagonal in any bay, the sum of which, be it remembered, should equal the total shearing force on the bay (see p. 16). For example, the vertical force taken by the top flange in bay No. 2, Diagram 3, is 10·3, which being added to 14·7 the vertical force taken by the diagonal gives a total of 25; an examination of the diagram will shew that $31 - 6$ or 25, is the shearing force on the bay.

In bay No. 4, of Diagrams 4 and 5, the vertical effect of the flange must be *subtracted* from that of the diagonal to get the true shearing force on the bay, since the direction of flange and diagonal with reference to the horizontal is *similar*, and their signs are opposite. Thus the flange of bay No. 4, Diagram 4, tends to push point l upwards with a force of 2·84, and the diagonal to pull it downwards with the same force, therefore the shearing force on the bay is 0.

In practice it would not be necessary to find the strains on every part of every bay as is done in Diagrams 1 to 6, but only in those bays which immediately adjoin the point where the live load changes from the dead load *only*, to dead and live load *combined*. Reference to Diagrams 1 to 6 will shew that it is in these bays that the maximum strains are to be looked for.

Comparison with a parallel Flanged Girder. For the sake of comparison, the strains on a parallel flanged girder whose depth is the same as that of the hogbacked girder at the centre, and whose loading is similar, are given in fig. 30.

Fig. 30.

It will be observed that the strains on the flanges of the hogbacked girder are in general the greater, and the strains on the diagonals the lesser of the two.

All the verticals in the parallel girder are in compression, whereas on the hogbacked some are in tension, and others have to resist both tension and compression.

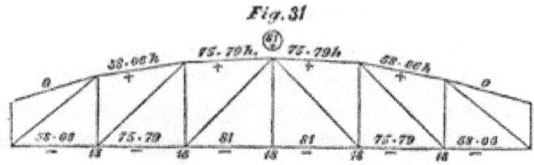
Fig. 31

Diagonals as Struts. Fig 31 represents a truss of the same dimensions and loading as that illustrated in Plate V., but having the position of the diagonals reversed, they being struts in this case. The same process may be adopted in this case as when tension-bars are used for calculating the strains on the diagonals.

Diagonals of both kinds used. When it is desirable to make use of diagonals both as struts and ties, or when a web of plate iron is employed, the most convenient plan will be to make the horizontal strains on the top and bottom flanges of each bay alike by taking a mean.

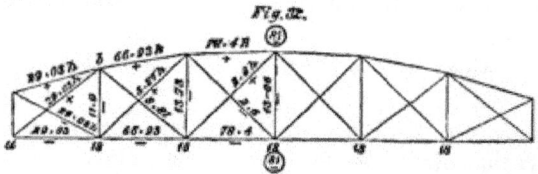
Fig. 32.

Fig. 32 represents a truss with equal horizonal strains on the top and bottom flanges of each bay. Examination of the diagram will shew that this necessitates equal horizontal strains on the diagonals in each bay, and that the verticals are in tension.

CHAPTER IV.

THE BOWSTRING GIRDER

Is so named from its resemblance to a strung bow. In all the various forms of girder that we have hitherto considered, the thrust of the top flange was conveyed to the bottom flange by means of a web or by struts and ties. The peculiar feature of the bowstring girder is that the thrust of the top flange bears directly against the end of the tie. In fact, the top member may be regarded as an arch, the thrust of which is taken by a horizontal tie instead of by abutments.

On p. 16, it is shewn that the strains on the flange of a horizontal parallel girder at any points are in the same proportion to each other as are the vertical lines drawn from these points to a parabolic arc, the apex of which is over the centre of the girder and whose extremities coincide with those of the flange. Since the strain on the flange of a girder is inversely proportional to the depth of the girder, it follows that by making the depth at every point along the girder proportionate to the vertical ordinate to the parabolic arc, we should obtain a girder in which the *horizontal* strain on the flanges was *equal* throughout. In fact, we should obtain a bowstring girder, the form of the bow being that of a parabola, the axis of which bisects the girder at right angles. When such a girder was uniformly loaded at the top the web would have (practically) no duty to perform and might be dispensed with, but it is required to resist the effects of unequal loading.

Curve of Equilibrium. A parabola is the *curve of equilibrium* for a uniformly distributed load, that is to say, the line of stress passes entirely along the curved flange so that it has no tendency to bend.

A *curve of equilibrium* is simply a line of pressure whose direction is curved; the direction of the pressure at any point in the curve being tangential to the curve at that point. A chain hangs in a curve of equilibrium, forming what is called a *catenary* (from *catena*, a chain). If it were so loaded as to have the same weight per *horizontal* unit of length it would assume a parabolic curve; by loading it irregularly it might be made to assume an irregular curve.

Method of drawing the Curve of Equilibrium. As we shall have to make use of the curve of equilibrium presently we must know how to draw it.

In most cases which come before us we shall find the load concentrated at intervals, so that our line of pressure will be polygonal instead of curved. We may however call the curved line which would pass through the angles of the polygon our curve of equilibrium.

Regular Loading. The simplest case will be that of a number of equal weights, horizontally equidistant, which have to be supported by a polygonal arched framing hinged at the joints. Let it be required to draw such a framing.

Let the span of the arch be divided up into six equal hori-

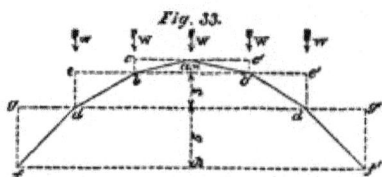

Fig. 33.

zontal bays, and at the junction of each bay let a load W act vertically. (See fig. 33.)

Then the load will be symmetrically distributed over the arch, one of the loads W being at the apex. Of this load half will go to each pier. Let ab and ab' be two beams meeting at a, making the same angle with the horizontal and supporting the load W. Each of these beams will support a vertical load of $\dfrac{W}{2}$, which may be represented by the vertical line bc or $b'c'$ drawn from the point

b or b' to meet the horizontal line drawn through point a in c or c'. The lines ac and ac' will represent the horizontal thrusts on the beams ab and ab', which are equal by construction, so that point a is in equilibrium.

At point b (dealing with one side of the polygon only, for the two sides must obviously be alike) an additional vertical force of W comes into play; thus the beam bd will have to carry a vertical load of $W + \dfrac{W}{2}$, or three times that on ab, and to resist the *horizontal* thrust of the beam ab. Therefore to find the position for the beam bd, draw from point b the horizontal line be, equal to ac; from point e let fall a perpendicular ed three times the length of the line bc; join bd: bd is the correct position for the beam.

Similarly, as the beam df has to carry a vertical load of $W + W + \dfrac{W}{2}$, the vertical gf must be made five times the length of cb, gd being made equal to ac. The points d' and f' on the other side of the polygon being found in the same manner, a curved line passing through the points $fdbab'd'f'$ would be called the curve of equilibrium. It is a parabola in this case, and would represent the line of pressure in an arch whose versed sine was ah caused by an evenly distributed load.

To draw the Curve of Equilibrium with a given versine. It is obvious that the curve $fdbab'd'f'$ is not the only form of arched polygonal framing that would support the weights W.

The angle at which the beams ab, ab' are inclined with the horizontal is arbitrary. The lines cb, ed and gf are in the proportions of the numbers 1, 3 and 5 respectively, and provided these proportions are preserved the lines may be of any length. That is to say, the versine of the curve may be made of any required dimension. The line ah representing the versine of the arch is equal to the sum of the lines cb, ed and gf, or $= 1 + 3 + 5 = 9$. We have only therefore to divide the versine in the proportion of 1, 3 and 5, as shewn in fig. 33, and to draw horizontal lines from the points so obtained to the corresponding vertical lines of action of the load and trace the curve through these points.

It will be observed that when one of the loads comes at the centre of the span, that is, when the number of bays is *even*, the versine or rise of the arch, as we may in fact call it, will have to be divided up into lengths which are terms of an arithmetical series $1 + 3 + 5 + 7 +$ &c. according to the number of bays. Also

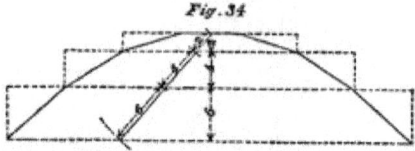

Fig. 34

that when the number of bays is odd the versine will be represented by the sum of a series $2 + 4 + 6 +$ &c. (see fig. 34), and in each case the last term of the series is *one less than the number of bays* in the span.

If $n =$ the number of bays in the span,

$\left(\dfrac{n}{2}\right)^2 =$ the sum of the series for a girder with an *even* number of bays,

and

$\left(\dfrac{n}{2}\right)^2 - \dfrac{1}{4} =$ .... *uneven* ....

The most convenient practical way to set out the curve is as follows. Take with the compasses from any convenient scale a length representing the sum of the series answering to the versine of the arch, and with the crown of the arch as a centre and this length as radius, describe an arc cutting the chord of the arch (see fig. 34): join this point of intersection with the centre from which the arc was struck, and mark off along this line divisions corresponding to the several terms of the series, commencing with the first term at the crown of the arch. Through the points so obtained draw horizontal lines to meet their corresponding verticals; the points of meeting will be in the curve of equilibrium.

Irregular loading, bays equal. It is obvious that the curve will be of the same form whether W the load per bay be great or small, but if the value of W be not the same for every bay the curve will be altered accordingly.

To draw the curve of equilibrium when the loads on the bays are irregular, the first step is to ascertain the total load which

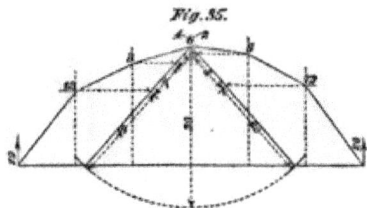

Fig. 35.

goes to each abutment. Fig. 35 represents the irregularly loaded framing: in this case the load on the left-hand abutment is 19 and on the right-hand abutment 20, therefore we must take 4 of the load of 6 at the crown to the left and 2 to the right hand.

The vertical loads supported by the beams
on the *left*-hand side will be respectively 4, 4 + 3 and 4 + 3 + 12,
„   *right*   „   „   „   2, 2 + 6 and 2 + 6 + 12,
that is   4, 7 and 19, sum = 30,
and   2, 8 and 20, „ = 30.

Taking the point at the crown where the load 6 divides as centre, describe an arc with a radius = 30 cutting the chord of the arch in two places; join each of these points of intersection with the centre, mark off divisions equal to 4, 7 and 19 on the left-hand line and divisions equal to 2, 8 and 20 on the right-hand line. Draw horizontal lines through the points of division to meet their respective verticals on either hand, as shewn in the figure; the points so obtained are the hinges of the framing, the beams of which coincide with the lines of thrust.

Irregular Loading, Bays unequal. It will be observed that the sums of the two irregular series are equal, being 30 for both sides. This corresponds to the equality of moment at the centre of the forces which go to each abutment, as previously pointed out in treating of irregularly loaded parallel girders (see p. 34). This correspondence will afford a convenient means of setting out the polygon of equilibrium when the bays are of unequal lengths.

Fig. 36 represents a polygonal framing in which the bays and loads are both unequal.

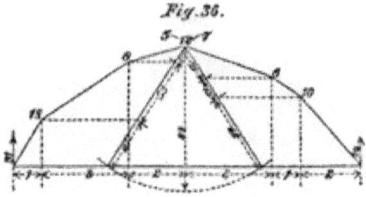

Fig. 36.

The moments of the loads which go to the abutments are as follows.

Left-hand abutment.					Right-hand abutment.				
loads	sums	leverage		moment	loads	sums	leverage		moment
$5+8+18=$	31	× 1	=	31	$7+6+10=$	23	× 2	=	46
$5+8=$	13	× 3	=	39	$7+6=$	13	× 1	=	13
	5	× 2	=	10		7	× 3	=	21
				80					80

That the above table of moments may be seen at a glance to be correct, we give in fig. 37 a diagram of a parallel girder loaded

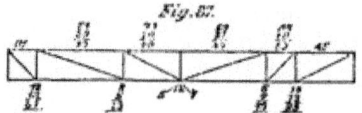

Fig. 37.

similarly to the framing, the additional horizontal moment on each several bay agreeing with that given in the tables. Now the *horizontal* strain throughout our polygonal framing must be the same for every beam; consequently the *depth* of the arch or vertical distance from any hinge to the chord must vary directly as the strain on the flange of the parallel girder at the point corresponding to the hinge. Therefore taking the versine of the arch as equal to 80, the heights of the first and second hinges to the left above the chord will be 31 and 70 respectively, on the right they will be 46 and 59.

By describing an arc with a radius = 80, having its centre at the hinge where the load divides, and cutting the chord in two places, and proceeding in a manner similar to that described on p. 49 for an irregularly loaded framing with equal bays, the positions of the hinges may be found.

Line of greatest resistance. Though in the preceding examples the quasi curve obtained was *the true* curve of equilibrium for the conditions given, since it must of necessity pass through the centre of each hinge-joint to preserve equilibrium, yet it does not follow that *a* curve of equilibrium drawn within an arch or curved rib will be *the* curve of equilibrium for that case. The line of pressure must always be the line of greatest resistance, its direction will therefore greatly depend upon the material as well as the form of the arch or curved rib. The true curve of equilibrium in an arch is the line of pressure which strains the arch *least*, so that if we draw the curve of equilibrium for an arch, and can shew that, assuming the pressure to pass along this curve, the arch is not overstrained, we may consider the arch as safe.

Methods of resisting Distortion. When a bowstring girder with a parabolic top flange is loaded evenly all over, the curve of equilibrium passes along the centre of the flange, and there is no tendency to distortion; but when the girder is loaded on one side only, the curve passes outside the bow on the one side of the span and inside on the other, if the flange be narrow, or close to the outer edge on one side and inner on the other if the bow be of considerable depth, as when it forms a curved box beam.

The distorting action on the bow may be resisted,

1. By making the bow sufficiently deep to include within its boundaries the curve of equilibrium under every condition of loading, it being wholly in compression.

2. By the transverse strength of the bow.

3. By the transverse strength of the tie.

4. By the transverse strength of the bow and tie combined.

5. By means of a web or diagonals.

First Method. Fig. 38 illustrates the first method. The dotted line touching the boundaries of the top flange or bow at *a, b* and *c* is the curve of equilibrium under one condition of unequal loading, say one half span fully loaded. (In describing the curve of equilibrium for this and the following cases we must

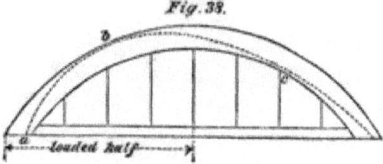
Fig. 38.

endeavour to draw it so that its maximum or minimum distance from the outer and inner edges of the bow is the same, as this is very near the truth, and gives the most economical results. It will be almost impossible in drawing the curve to hit off on the first occasion its true position; but by lengthening or shortening the ordinates to the curve *in the same proportion*, we can raise or depress the curve until its position is approximately correct.) In Plate VI. the strain on the top flange of a certain truss fully loaded is given as 81 at the centre; when the same truss is only half loaded, the strain at the centre is given as 60·75. Supposing our bowstring girder to have the same span, depth at centre, and conditions of loading as the truss in Plate VI., there would be a *horizontal* strain of 60·75 at points a, b and c. By changing the position of the load we can cause the curve of equilibrium to pass very near to almost every point in the outer and inner edges of the bow, and at all of these points the strain would have a horizontal element of 60·75, or thereabouts. Thus it will be seen that metal will have to be provided to take a horizontal thrust of 60·75 throughout the outer and inner edges of the bow, or 121·5 in all, whereas had the curve of equilibrium passed along the centre of the flange, a strain of but 81 would have to be provided for. The curve of equilibrium may however fall considerably within the

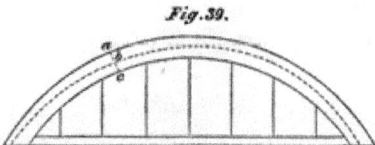
Fig. 39.

edge of the top flange; in such case the compressive force on either edge will be inversely proportional to its distance from the line of pressure, the *sum* of the horizontal strains on the two edges being equal to the thrust on the tie; for example,

S being the total force acting along the dotted line at b, fig. 39,

$$\text{Strain at } a = \frac{bc}{ac} S,$$

$$\text{„ „ } c = \frac{ab}{ac} S.$$

Second method. In fig. 40, the curve of equilibrium shewn by the dotted line passes outside the limits of the top flange

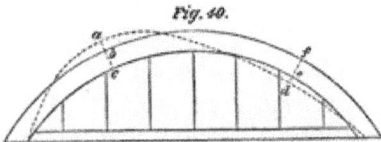

Fig. 40.

in two places, the extreme distances being ab in the one case and de in the other. With the curve in this position, the tendency of the bow is to flatten at b and bulge outwards at e.

S being the pressure acting along the dotted line,

Strain on inside edge at $c = \dfrac{ab}{bc} S$ tension, b being the fulcrum,

„ outside „ $b = \dfrac{ac}{bc} S$ compression, c „ „

„ inside „ $e = \dfrac{df}{ef} S$ tension, f „ „

„ outside „ $f = \dfrac{de}{ef} S$ compression, e „ „

It must be remembered that in this and in the last example, S is not a uniform force, it will vary with the angle of inclination of the curve of equilibrium with the horizontal. The *horizontal element* of S *is* uniform.

Third method. Fig. 41 represents a bowstring girder in which the bow is quite flexible, that is hinged at the points where it joins the verticals. The tie affords the means of resisting distortion. When the left-hand side of the span is loaded, the

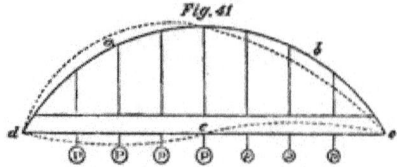

Fig. 41

arch tends to flatten at a and bulge outwards at b, causing the horizontal tie to bend downwards between d and c, and upwards between c and e. It assumes in fact the form of an S curve as shewn by the dotted line d, c, e, the point of contrary flexure being at or very near to the centre of the span.

To prevent the top flange from buckling it is evident that the bottom flange must have sufficient transverse strength to resist the bending action to which it is subject. Let us endeavour to ascertain what this should be.

Let us suppose the left-hand half of the span to be loaded, and the load to be suspended from the bottom tie, this will throw all the verticals into tension. Now if the bottom tie were flexible and the bow stiff as in the immediately preceding cases, the strain on the verticals on the left would be $P = $ dead + live load, while on those to the right it would be $p = $ dead load. But as our parabolic arc is perfectly flexible and is the curve of equilibrium for an evenly distributed load, it is evident that in order to preserve its form the strain on each vertical must be the same throughout the span, it will therefore be $\dfrac{P+p}{2}$. But since a load of P is attached to the bottom flange underneath each vertical on the left-hand half of the span of which an amount $= \dfrac{P+p}{2}$ is carried by the vertical, the remainder

$$P - \frac{P+p}{2} = \frac{P-p}{2}$$

must be supported by the transverse strength of the tie offering resistance to a vertical *downward* force. Again, the upward pull of $\frac{P+p}{2}$ on the verticals of the left-hand side of the span is met by a load of p only, therefore that part of the tie is subject to a vertical *upward* force of

$$\frac{P+p}{2} - p = \frac{P-p}{2}.$$

Now $P-p$ = the live load on each bay, therefore the loaded half dc of the tie is subject to a load = half the live load per bay acting downwards, and the other half of the tie ce to the same force acting upwards. When the position of the load is reversed the directions of the transverse strains on these half-lengths of horizontal tie will also be reversed, so that *each half of the tie must be made capable of resisting an upward and downward transverse force equal to one half the live load on half the span,* or *one quarter the whole live load on the span*[1]. The tie may be hinged at c the centre, if necessary, without impairing its effect in resisting distortion. Each half of the tie must be treated as a distinct girder of a length equal to half the whole span of the bridge. Since the tie is in tension throughout all its cross-section, the transverse strain, unless very great indeed, will not put any part of the tie into compression. Thus if the horizontal strain on the tie, when the bridge is loaded over one half of its length only, be 100 or 50 on each flange, neglecting web (see fig. 42), and the strains due to transverse downward force be 10 + in the top flange and 10 − in the bottom flange, the addi-

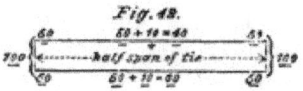

Fig. 42.

tion of these strains will give 40 − for the top flange and 60 − for the bottom: when the transverse force acts upwards the strains will be 60 − in the top flange and 40 − in the bottom flange. It would therefore be necessary to insure rigidity to increase the sectional area of both flanges at the centre of each half span. The extra metal may be diminished to nothing at the centre and ends of the span of the arch.

[1] The "live load" here mentioned means that which comes upon *each girder*, which will of course be $\frac{1}{2}$ the whole load on the span if there are only two trusses.

Fourth Method. Bowstring girders are often made with a deep trough-like bow and tie; by utilizing the transverse stiffness of *both* these, resistance to distortion can be obtained.

Let us assume that it is desired to divide the work *equally* between the top and bottom members of the girder. This as regards the tie is equivalent to reducing the live load by one half in the preceding case, and proceeding as before. For the bow, we must draw the curve of equilibrium for one side only loaded, but the total load at each vertical of the loaded side must be taken as $= \text{dead load per bay} + \dfrac{\text{live load per bay}}{2}$. Having correctly drawn the curve the strain on the flanges of the bow can easily be determined, as explained on p. 53.

Fifth Method. The general method of providing against the distorting effect of unequal loading is by means of a web or cross-bracing between the bow and the tie.

Plate VI. contains a series of Diagrams of the strains on every part of a bowstring girder under unequal loading. It is necessary to find the strains on the flanges first and thence to deduce the strains on the diagonals, as in the case of the hogbacked girder. The mode of procedure is precisely the same as that adopted for the hogbacked girder; see p. 41. The depth at centre, number and width of bays and loading are the same as those taken for the hogbacked girder, Plate V., so that the strains on the two may be compared.

Effect of elasticity in distributing the Load among the Diagonals. As we have the bow of our girder in the form of a parabolic curve, when the load is uniformly distributed the curve of equilibrium will pass along the centre line of the top flange, the strain on the tie will be the same throughout its whole length, and there is no need of diagonals to prevent distortion. Diagram, No. 1, Plate VI. represents the girder under this condition of things. In this Diagram the diagonals are wholly omitted, and the load is assumed to be carried by the vertical rods only. In reality the strains are not such as are here given, because in consequence of *the stretching* of the verticals a strain comes upon the diagonals.

THE BOWSTRING GIRDER.

The accompanying figure represents a pair of diagonals inclined at an angle of 45° with the vertical and a vertical at whose point of meeting a a load is suddenly applied. The effect is to bring the point a down to b, stretching the vertical to the extent ab and bringing the diagonals into the position shewn by the dotted lines. Taking the upper extremity of one of the diagonals as a centre, and describing an arc through the point a cutting the dotted line in c, it becomes evident that bc represents the amount of stretching of the diagonal. Now since the distance ab is very small, the triangle acb is practically a right-angled triangle in which ab is the hypothenuse and ac and cb are equal sides. And the line ab is to the line bc as $\sqrt{2}$ (or 1·41 nearly) is to 1.

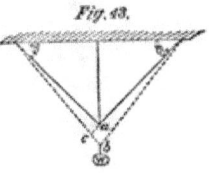

Fig. 43.

Let $ab = \dfrac{1}{100}$ th of the length of the vertical before it was loaded, the proportion being expressed by the fraction

$$\frac{100}{1} = \frac{\text{length of vertical}}{\text{amount of its extension}}.$$

Now, diagonal : vertical :: $\sqrt{2}$: 1, and the amount of the diagonal's extension expressed in terms of ab

$$= \frac{bc}{ab} = \frac{1}{\sqrt{2}};$$

$$\therefore \frac{\text{length of diagonal}}{\text{amount of its extension}} = \frac{100 \times \sqrt{2}}{\dfrac{1}{\sqrt{2}}}$$

$$= \frac{\sqrt{2} \times \sqrt{2} \times 100}{1} = \frac{200}{1}.$$

From this it appears that if each of the diagonals and the vertical be of the *same sectional area*, the *sum* of the strains on the two diagonals will be as nearly as possible equal to that on the vertical.

In this case the strain on the diagonals is 1·41 times the load supported by them, therefore if

$x =$ proportion of load taken by vertical,

$y =$,, ,, ,, ,, diagonals,

$x + y = W$ total load carried;

but $\quad x = 1{\cdot}41\, y$;

$\therefore 2{\cdot}41\, y = W$,

$$y = \frac{W}{2 \cdot 41}.$$

If $W = 1$, $y = {\cdot}4149$ and $x = {\cdot}5851$.

These values of x and y are of course only true when the diagonals are inclined at an angle of 45° with the vertical and the sectional area of each diagonal is equal to that of the vertical. To find the proportions of load taken by diagonals and vertical or any system of bars, the general rule is:—Find the proportion of extension to its length which each bar experiences, assuming the load to descend vertically through *a given space*. The strain *per square inch* thus placed upon the bar multiplied by its area in *square inches* gives the actual strain upon that bar. Finally, ascertain the vertical element of the strain so obtained. This being done for each bar the share of the load carried by each will be *in proportion* to the vertical strain upon it.

For example. The three bars *ab, ac, ad* (fig. 44) support a vertical load of 100 between them:

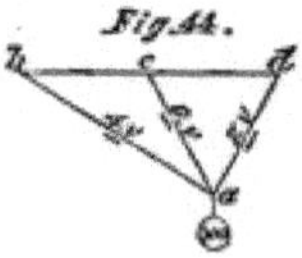

parts

ab supporting $4 = \dfrac{4 \times 100}{20} = 20$ actual load carried.

ac ,, ,, $9 = \dfrac{9 \times 100}{20} = 45$,, ,, ,,

ad ,, ,, $7 = \dfrac{7 \times 100}{20} = 35$,, ,, ,,

Total $\overline{20}$ $\quad\quad\overline{100}$

Fig. 44.

In dealing with the diagonals of a truss it must be remembered that the extension of the bars may be effected by the yielding of their supports at the upper extremities. Thus if *b, c, d*, fig. 44, represent the top flange of a truss, its compression under load would cause the points b, c and d to approach one another, and thus affect the lengths of the bars ab, ac, ad. Further, if the points b, c and d were supported as to vertical motion by struts, their relative depressions would depend upon the strength of the struts which supported them, and again upon the strength of the ties which supported these struts, and so on *ad infinitum*.

Thus it is found practically impossible to tell the exact strain upon a number of bars in a truss inclined at different angles and meeting at a point where a load is applied.

The usual practice is to calculate the strain on every bar under the most unfavourable condition and to neglect the easing effect upon it of other bars, the strain upon which may be much or little, but is uncertain.

The strains on Diagrams 1 to 6, Plate VI., are strictly correct for trusses of the form represented in those Diagrams, but they are not true as regards the verticals for one similar to Diagram 7, in which the diagonals cross one another. However, by taking the strains as given in Diagrams 1 to 6 as true for a truss of the form of Diagram No. 7, we shall be *safe*.

The reader will observe this peculiarity in the Diagrams for the bowstring girder, that the live load being removed from one end of the girder the diagonals which come into play slope all in the same direction. As a natural consequence the horizontal strain on the top members increases in each bay till it reaches its maximum in the last bay at the loaded end of the girder. If the diagonals were struts their slope would of course be reversed. The diagonals throughout may however be made capable of acting either as struts or ties. In such case the extreme strain on each single diagonal given in Diagram 7 would have to be divided between the two diagonals of the bay. This is analogous to the effect of a solid web capable of resisting compression and extension; see page 24.

Figs. 45 and 46 shew two other alternative methods of arrang-

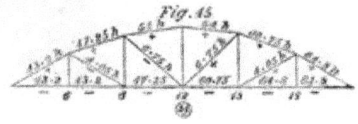

ing the diagonals which should be made capable of resisting both extension and compression. For the purpose of shewing this the

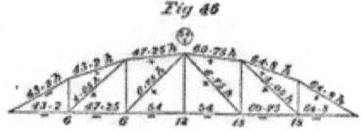

strains are given for each system, the girders being taken as half-loaded.

Although not to be commended for appearance, trusses with *single* double-acting diagonals are cheap, and have this advantage, that the strains can be predicted with certainty.

Load resting on the top. The effect of resting the load upon the top flange instead of suspending it from the bottom is to

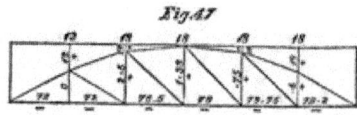

alter the strains upon the verticals only, they also becoming struts. Fig. 47 corresponds to Diagram No. 2, Plate VI., but gives the strains when the load rests upon the top of the girder. With this example before him, the student should be able to construct without difficulty the Diagrams for the other positions of the load.

The curve of the bow is often made an arc of a circle for appearance sake and convenience of manufacture.

CHAPTER V.

THE ARCH.

The preceding Chapter contains much that we shall have to refer to in treating of the Arch in consequence of its affinity to the bowstring girder as indicated in the opening paragraph of that chapter.

Method of finding the Thrust at Crown. The horizontal thrust on the crown of an arch may be found in the following manner.

Consider the arch with its spandrils, roadway, and live load, as forming two rigid masses divided vertically at c the crown of the arch, and hinged at a the the centre of the skew-back, (see fig. 48). Find the centre of gravity of the mass a, b, c, being one-half of the whole arch. Draw a vertical line through this point to represent the line of action of the weight which we will call $\frac{W}{2}$. This acts at an arm of z, about the fulcrum a. Draw a horizontal line cb through the centre c of the arch to represent the line of action of the thrust. Through a, draw ab perpendicular to cb, the line ab represents the leverage at which the thrust acts about point a, and as this thrust is the force that balances $\frac{W}{2}$;

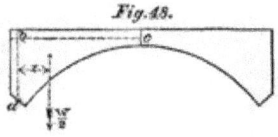

Fig. 48.

$$\therefore S \times ab = \frac{W}{2} \times z, \text{ if } S = \text{thrust,}$$

and
$$S = \frac{W \times z}{2ab} \quad \ldots\ldots\ldots\ldots\ldots\ldots\ldots\ldots (1).$$

Curve of equilibrium for an Arch of Masonry. An arch of masonry is stable when the curve of equilibrium lies well within the arch under all conditions of loading. The mode of drawing the curve of equilibrium has been fully described in the preceding chapter. An example of the method of procedure in the case of an arch of masonry is given in fig. 49. The arch should be

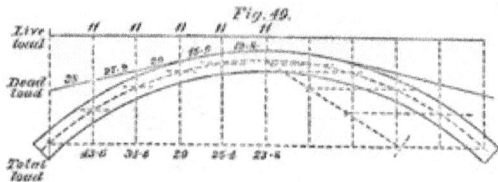

divided up into equal horizontal segments; the weight of each portion should be calculated, and the mean weight of two adjoining segments considered as acting in the direction of the vertical line which divides them. To the dead load must be added the live load when necessary. For the loads so obtained a curve of equilibrium must be drawn through those points in the crown and in the haunch of the arch which appear likely to give the best results, which are obtained when the distance of the curve from the outer or inner face of the arch is a maximum. The available area of masonry for resisting thrust is equal in width to *twice the minimum distance* of the line of pressure from the extrados or intrados of the arch. If the points through which the curve has been drawn have been injudiciously chosen, the position of the curve can be changed by altering all the ordinates in the same proportion.

Other methods of resisting distortion of the Arch. In the case of metal ribs, the resistance to distortion may be effected as in the cases of the bow-string girder by the five methods mentioned on p. 51, substituting the word *rib* for *bow*, *horizontal girder* for *tie*, and adding *in the spandrils* after the word *diagonals* to the description of the 5th method. The first four methods are exactly analogous for the bowstring girder and the arched rib, so that it is unnecessary to say more about them, but the method of finding the strains on the diagonals in the spandril of an arch differs from that used for the bowstring girder.

Strains on the Spandrils. Diagrams 1 to 6, Plate VII., shew the strains caused upon the various parts of a parabolic arched rib, whose spandrils are braced with a single diagonal, under various conditions of loading. Diagram 7 gives the greatest strains collected from Diagrams 1—6. The form of the arch, and conditions of loading, are the same as those of the bowstring truss illustrated in Plate VI. In Diagram No. 1 are given the strains caused by the full load. The arch being parabolic, there is no strain on the diagonals under this condition of loading. To find the strains when the load is unequally distributed, and the diagonals are placed as in Plate VII., the most convenient method is to commence at the springing of the arch, and work towards the centre.

We will now deal with Diagram, No. 2, Plate VII., commencing at the left-hand corner.

In the first place, we find the vertical load on the abutments which amount to 40 for the left-hand, and 44 for the right-hand abutment (see p. 33). In the next place, we find the thrust of the arch on the principle mentioned on p. 61. Thus

```
load  leverage  moment
 12  ×    1   =   12
 18  ×    2   =   36
 10  ×    3   =   30
                  ──
                  78 horizontal thrust.
```

Of course the thrust on the right-hand abutment is the same. Thus

```
load  leverage  moment
 18  ×    1   =   18
 18  ×    2   =   36
  8  ×    3   =   24
                  ──
                  78 horizontal thrust.
```

Now the segment of the arch forming the lower member of bay No. 1 being inclined at the rate of 5 vertical to 9 horizontal, the vertical pressure which it exerts is to its horizontal as 5 is to 9,

$$\therefore 78 \times \frac{5}{9} = 43\cdot 33,$$

the vertical pressure.

But since there can only be a vertical pressure of 40 on the abutment, the vertical bar which meets the rib at the springing must exert an upward pull of $43 \cdot 33 - 40 = 3 \cdot 33$ tension.

From this we derive a vertical thrust of $3 \cdot 33$ on the diagonal of bay No. 1, which answers to a horizontal of $7 \cdot 5$, which again requires a strain of $7 \cdot 5$ in tension on the top member of bay No. 1. To maintain equilibrium, there must be a horizontal thrust of $78 + 7 \cdot 5 = 85 \cdot 5$ by the rib in bay No. 2. The vertical element being $\frac{3}{9}$ of this $= 28 \cdot 5$.

The rib in bay No. 1 is pressing point a upwards with a force of $43 \cdot 33$, while the rib in bay No. 2 and the diagonal in bay No. 1 are pressing it downwards with a force of $28 \cdot 5 + 3 \cdot 33 = 31 \cdot 83$,

therefore $\qquad 43 \cdot 33 - 31 \cdot 83 = 11 \cdot 5$

is the pressure which the vertical bar must exert to maintain equilibrium. Therefore of the load of 12 which is supported at the upper extremity of this vertical $\cdot 5$ will be carried by the diagonal of bay No. 2. This bar will exert a horizontal thrust of $4 \cdot 5$, which will cause the total horizontal strain on the top member in bay No. 2 to be -12; this strain will be carried through to bay No. 3.

Now $85 \cdot 5 + 4 \cdot 5 = 90$, the combined horizontal thrust of rib and diagonal in bay No. 2, which must be met by a thrust of 90 on the rib in bay No. 3. The vertical element of the thrust on this part of the rib will be $90 \times \frac{1}{9} = 10$, and 10 is the proportion of the load of 18 at the centre which goes to the left.

Again, the point b is pressed upwards with a force of $28 \cdot 5$ by the rib in bay No. 2 and downwards with a force of $\cdot 5$ by the diagonal of bay No. 2, of 10 by the rib in bay No. 3, and of 18 by the vertical bar dividing the bays; the total downward force being $28 \cdot 5$.

Further, $90 - 12 = 78$ the thrust at the crown, which agrees with that at the haunches. We find therefore, that the conditions of equilibrium are satisfied throughout, and we may conclude that our strains are correct.

The strains on the right-hand side of the arch may be obtained by a similar mode of proceeding. In this part of the arch the diagonals of the spandril are in tension.

For the arrangement of diagonals given in Plate VIII., the most convenient way of calculating the strains is to commence from the centre of the arch.

For the arrangement of loading given in Diagram No. 2, Plate VIII., we know that of the load 18 resting at the crown 10 goes to the left, consequently there must be a vertical load of 10 on the rib in bay No. 3; this will produce a horizontal thrust of 90 upon it, which will produce a horizontal thrust also of 90 on the rib in bay No. 2 of which the vertical thrust will be 30, necessitating a downward pressure of 20 by the vertical at point b. To produce this, the diagonal in bay No. 2 must exert a downward pull of 2, and therefore a horizontal pull of 4·5. Thence we determine the total horizontal thrust on point a to be $90 - 4\cdot5 = 85\cdot5$, which must be the thrust on the rib in bay No. 1. The vertical pressure exerted by this part of the rib will be 47·5, necessitating an upward pull of 7·5 by the diagonal in bay No. 1, to reduce the total load on the abutment to 40. This diagonal will exert a horizontal pull of 7·5 also on the top member of bay No. 2, and cause the total load on the vertical at a to be 19·5.

The diagonal in bay No. 2 causes the downward pressure of the vertical at b to amount to 20, and adds 4·5 of horizontal strain to the top member, making a total of 12 tension thereupon.

Comparing the corresponding diagrams of Plates VII. and VIII., we find that the *horizontal* strains on the diagonals are *the same* for *corresponding bays* under similar conditions of loading.

In Diagram No. 8, Plate VIII., are given the greatest *horizontal* strains on diagonals, rib, and top member, when the diagonals forming the spandrils are crossed. Diagram No. 8 is the *sum* of Diagrams 7 in Plates VII. and VIII. It also gives the greatest strain on the verticals. The load is 36 per bay for this diagram; for a load of 18 it would be necessary to halve the amounts.

We observe, therefore, that in the spandrils of arches constructed on the system of Diagram 8, Plate VIII., the *horizontal* strain on each diagonal of a bay is the same. Wherefore it is sufficient, so far as the diagonals are concerned, to calculate the strains for *one* system only.

The two following facts are to be noted with regard to the Diagrams on Plates VII. and VIII.

1st. If in each bay of a Diagram the horizontal strains on the diagonal, top member, and rib, be added together, the − cancelling the +, the remaining + will be the same for each bay, and will equal the thrust of the arch at the crown.

2nd. The sum of the vertical strains on diagonal and rib in any bay, the − cancelling the +, is equal to the shearing force on that bay.

In illustration of the first statement, add together the horizontal strains on the severals bay of Diagram No. 2, Plate VII., and the resulting strain will be found to be + 78h.

Again, adding the vertical effect on each bay, we get

for bay No. 1......40
„   „   „   2......28
„   „   „   3......10
„   „   „   4...... 8
„   „   „   5......26
„   „   „   6......44.

These amounts are the correct shearing forces for their several bays.

The extreme strain which these diagrams shew on the rib in bays 3 and 4, would not arise in actual practice, for the horizontal girder and rib are united together for some distance on each side of the centre of the arch in most arched bridges; this being the case, the tension on the horizontal girder neutralizes some of the compression on the rib. If, however, the horizontal girder be at a higher level than the crown of the arch, the variation of strains would not be so great as it appears in the case of which we have just treated.

Other forms of Spandril filling. Figs. 50, 51, and 52, represent three other regular forms of spandril filling; for the first of these systems the strains may be obtained by proceeding as in the last example, making a separate series of diagrams for each system of diagonals. In figs. 51 and 52 the resistance to distortion is effected by the transverse stiffness of the rib and horizontal girder, (see p. 62).

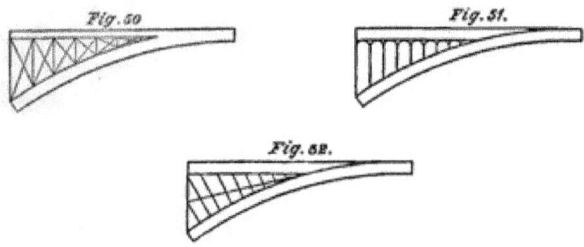

The inclined position of the spandril bars in fig. 52 will cause the curve of equilibrium for the strains on the rib to be an approximation to an arc of a circle when the bridge is fully loaded; indeed the angle of the bars might be so adjusted that the curve of equilibrium should be exactly an arc of a circle under the full load, and correspond with the centre of the rib. The effect of unequal loading in a bridge of this description on the horizontal girder would be to introduce an additional horizontal stretching or compressing besides transverse strain.

Arched bridges of multiple span. The preceding pages treat of the strains which come upon an arch of a *single span*. We have now to consider the strains which arise from unequal loading on an arched bridge of many spans.

When the piers of an arched viaduct are short, it may so happen that the system of arching may be stable without the aid of the spandrils.

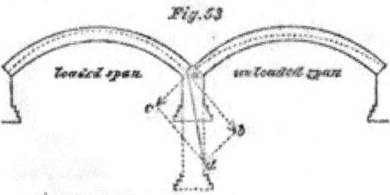

In fig. 53 are shewn two arches of a viaduct, one of which only is loaded. The curved dotted lines drawn within the

5—2

limits of the arch represent the curves of equilibrium for each arch. From the point of their intersection a, lines ab, ac are drawn tangential to the curves, and representing to scale the strain on the arches, ab being that on the loaded arch, and ac that on the unloaded arch. The parallelogram being completed, the diagonal ad represents the resulting line of pressure. In this example the line of thrust falls well within the base of the pier, so that the arching may be pronounced stable. The dotted prolongation of the pier and line of thrust below the ground-line, is intended to shew that if the pier were of the height shewn by the dotted lines, the line of thrust would fall outside the base and the arching be unstable.

Stability of Pier. The dead weight of the pier contributes to the stability of a system of arches, as fig. 54 will serve to shew. Let ab be the resultant thrust of the two arches, o the centre of gravity of the pier; through o draw a perpendicular to intersect the line of thrust in a, take $ac =$ the weight of the pier, complete the parallelogram $abcd$, join ad; ad is the resulting line of thrust which now falls within the base.

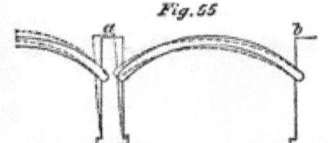

Effect of horizontal Girder. Figs. 55 and 56 are intended to shew the action of a horizontal girder in preserving the stability of an arch. In fig. 55, there being no horizontal girder, the weight of the pier is the sole source of stability; this being overcome, the upper ends of the piers a and b are forced apart. If these are tied together by means of a horizontal girder the arch can only yield by the pier breaking as at c, fig. 56.

Transverse strain on Pier. In estimating the transverse strain upon the pier at c, find the *surplus horizontal thrust* at c arising from the loaded span, consider the whole pier as a lever fastened at its two extremities and acted upon by a transverse

THE ARCH.

force at c, the breaking tendency can then easily be found. To find the value of the surplus horizontal force, resolve the thrusts of the two adjacent arches into their horizontal and vertical elements, consider the vertical effect of the thrusts as acting downwards through the point c (see fig. 57), the neutral axis of the pier at the springing, add to this the weight of that part of the pier above c, the point of rupture, and call the *whole weight* W. Again, call the weight of the lower part of the pier acting through its centre of gravity o, P. Choose a point in the base of the pier as a fulcrum or turning point.

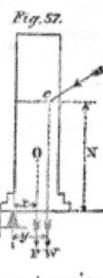

Fig. 57.

Let x be the leverage of P, y the leverage of W, and z the distance from base of pier to point of rupture.

Then $\dfrac{Px + Wy}{z}$ is the resistance which the dead weight offers to rupture.

Add this force to the horizontal thrust of the *unloaded* arch, and subtract the whole from the horizontal thrust of the loaded arch, the remainder is the *surplus horizontal thrust*.

It is manifest that the resistance of the horizontal girder in the *unloaded* span to *compression*, helps to prevent the pier from overturning.

In viaducts of masonry the spandril filling acts partly as a horizontal girder in preventing the overturning of the pier, and partly as bracing preventing alteration of form in the arch.

In dealing with a viaduct of masonry, and describing the curve of equilibrium for loaded and unloaded spans for the purpose of finding the resultant thrust, the curve in the loaded span should be carried as near to the extrados of the arch at the crown, and as close to the intrados at the springing as is consistent with safety; while for the unloaded span, the curve should be brought as close to the intrados at the crown as safety permits, and be made to intersect the curve of the loaded arch as high up as is possible without causing the resultant pressure to pass so near to the face of the masonry at the springing as to crush it. Fig. 53 illustrates this method.

THE ARCH.

Stability of a Wrought-iron Arched Viaduct. In a wrought-iron arched viaduct the question of stability is somewhat complicated.

Fig. 58 represents a wrought iron viaduct, perfectly continuous throughout. We might calculate its stability in the manner adopted for the case illustrated in fig. 56, and supposing it were found that the piers were too weak to ensure the stability of the system, this might be secured by making the rib stiff enough at the crown to resist the upward force which comes upon it. The arching must in that case be regarded as a continuous girder of varying depth in which the effect of the load is to flatten

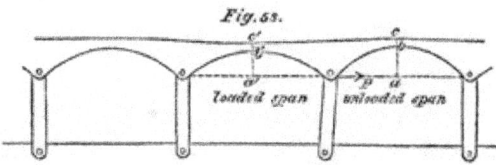

Fig. 58.

the loaded arch and to raise the crown of the unloaded arch. The piers must be assumed to be hinged at the springing and at the ground-level, so that no assistance is obtained from them, because we are now considering the sole effect of the metal at the crown in affording stability.

The first step in the calculation is to find the *surplus horizontal thrust* at the springing. (In this case there is no resistance afforded by the pier.) Call this force P, take *half* this force or $\dfrac{P}{2}$ as acting at a leverage of ab (the versine of the arch), fig. 58, tending to crush the bottom flange of the rib at b, and to tear it open at c.

Let S be the strain in tension at c caused by the force $\dfrac{P}{2}$, then

$$S \times bc = \frac{P}{2} \times ab;$$

$$\therefore S = \frac{P \cdot ab}{2bc}.$$

The compressive strain at b will be $\dfrac{P}{2} + S$,

or
$$\frac{P}{2}\left(1 + \frac{ab}{ac}\right).$$

We take the horizontal force tending to force up the crown of the unloaded arch as $\frac{P}{2}$ only, because half the force P is absorbed in flattening the loaded arch.

On the loaded arch the force $\frac{P}{2}$ acts at an arm of $a'b'$ also; therefore if S' be the strain in *compression* on the top flange caused by the force $\frac{P}{2}$,

$$S' \times b'c' = \frac{P}{2} \times a'b';$$

$$\therefore S' = \frac{P \cdot a'b'}{2 \cdot b'c'}.$$

The tensive strain at b' will be $\frac{P}{2} + S'$,

$$\text{or } \frac{P}{2}\left(1 + \frac{a'b'}{b'c'}\right).$$

These strains must be added to the strains which result from the position of the curve of equilibrium to obtain the extreme strain upon the flanges.

In practice the stability of the pier must be taken into account. Indeed in the majority of cases it will be on the pier that we must rely: for before any considerable strain can be produced on the rib in the unloaded span, the span of the loaded arch must have increased through the forcing apart of the piers, as shewn in fig. 58. The amount of this alteration in length of the span will vary according to the height, thickness, and material of the pier. By measuring the actual increase in length of the loaded span, it is possible to determine how much of the resistance to overturning is offered by the stability of the piers, and how much by the stiffness of the ribs. This cannot however be done without a knowledge of the laws of deflection.

CHAPTER VI.

SUSPENSION BRIDGES.

A SUSPENSION bridge is merely an arched bridge inverted, the chain of the suspension bridge being subject to a strain in tension corresponding to the strain in compression on the rib of the arch.

There is one important practical difference between the two: this, that in the suspension bridge the chain being flexible is able to adapt itself to the curve of equilibrium, whereas no such property belongs to the arched rib.

This peculiarity of the suspension bridge causes the platform of the bridge to alter its form under the passage of a rolling load, unless the bridge be rendered rigid by some kind of bracing.

Strain of centre of Span. The strain on the chains at the centre of the span may ordinarily be found by the formula $\frac{WL}{8d}$, since the weight of the chains is small compared with that of the platform and live load, and consequently the load may be taken as *evenly* distributed. In this formula

$W =$ the total load on the bridge.

$L =$ the span between towers.

$d =$ the depression of the chain at centre below the points of support on the towers.

In very large bridges, however, where the weight of the chains is relatively large, this formula will not give correct results, and it will be necessary to describe the curve of equilibrium by the method described on p. 46, in order to obtain the strains on the chain at the centre of the span.

In ordinary cases the chains will hang in a curve which is a close approximation to a parabola. In cases where the chains are relatively heavy, the curve will approach more nearly the form of a catenary.

Strain at any portion of the Chain. The strain at the centre of the span being known, the strain at any point a (fig. 60) in the chain may be found by drawing a tangent to the curve at that point, cutting at b the horizontal line which is the tangent to the lowest point in the curve. If θ be the angle that the tangent makes with the horizontal, and S the strain on the chain where it is horizontal; then

Fig. 60.

$$\frac{S}{\cos \theta}, \text{ or } S.\sec \theta, \text{ is the strain at } a.$$

Or the strain may be found by geometrical construction, thus: let fall from point a a perpendicular to meet the horizontal tangent in c.

Then, strain at $a : S :: ab : bc$;

$$\therefore \text{ strain at } a = S\frac{ab}{bc}.$$

In the triangle abc, ab represents the actual strain at a.

bc „ horizontal element $= S$.

and ac „ vertical element of the strain
$=$ weight of superstructure between c and centre of span.

Strain on the back tie. As in the arch, the thrust of the arch is taken by abutments, so in the suspension bridge the pull of the chains is resisted by back ties, which are in general anchored to masses of masonry at a considerable depth below the surface of the ground in the abutments of the bridge. The strain on the back tie is found by means of a parallelogram of which the adjacent sides are tangents to the chain at its junction with the saddle on the pier and the back tie. The strain on the chain being known, that on the back tie is obtained by completing the parallelogram, the vertical diagonal of which represents the load on the tower.

In those suspension bridges in which no means are provided for resisting the distorting effect of unequal loading, the curve of the chain alters its form during the passage of the moving load.

Usually the moving load is small compared with the dead load, and consequently the alteration in the form of the chain is not of serious moment. Such bridges are, however, liable to accident from the undulations produced by high winds.

Rigidity can be obtained by means similar to those which are applicable to the bowstring girder as given on p. 51.

The adoption of the 1st, 2nd or 4th of these systems necessitates the substitution of a curved rib for a flexible chain, thus producing what has been called "an inverted arch bridge."

Whatever system of stiffening may be adopted, the whole structure should be hinged at the centre of the span, to allow for the rise and fall of the chain caused by variations of temperature.

The methods of finding the strains on the various parts of the suspension bridge are precisely similar to those already given for finding the strains on the corresponding parts of the bowstring girder, to which the reader is referred.

The usual practice is to stiffen the suspension bridge by means of a horizontal girder. The method of calculating the necessary strength of this girder is given on pp. 54, 55, and 56. The horizontal girder answers to the tie in the bowstring girder, but is not like it subject to longitudinal strain (see p. 55). It acts *only* as a stiffener. The ends on the abutments should be anchored, but should be free to move horizontally to allow for change of temperature.

To restate the rule, in a form applicable to the suspension bridge. If W be the whole live load on the bridge, then the horizontal girder, which reaches from the abutment to the centre of the span, will, under the most favourable conditions of loading, be subject to an upward and downward transverse force equal $\dfrac{W}{8}$ distributed.

If l be the span of the bridge,

d ,, ,, depth of the horizontal girder,

S the strain on the flange at centre,

$$S = \frac{\frac{W}{8} \times \frac{l}{2}}{8 \cdot d} = \frac{W \cdot l}{128 \cdot d},$$

Plates VII. and VIII. turned upside down with the signs of the strains changed, give the strains on a suspension bridge stiffened by means of diagonal bracing. As under the condition of loading, given in Diagram 5, Plate VII. or VIII., the rib of the arch becomes subject to tension near the crown, so, under the same condition of loading, the suspension chains are subject to compression.

For further remarks on suspension bridges see supplementary Chapter on Economy in Suspension bridges.

CHAPTER VII.

DEFLECTION.

BEFORE investigating the properties of continuous girders, it is necessary to become acquainted with the laws of *Deflection*, which, in the continuous girder, very importantly determine the magnitudes and directions of the strains.

Deflection is due to elasticity.

Definition. The deflection of a beam or girder is its vertical displacement by a load from the position that it occupied when unloaded, its bearing points being fixed.

So that were the bottom flange of the girder horizontal in its unloaded state, when loaded it would be curved downwards, the distance between the lowest point in this curve and the original horizontal line being the measure of the deflection, generally called *the deflection* of the girder.

If the flanges of the girder be parallel, straight, and of uniformly proportioned strength, the curve which it would assume when loaded would be an arc of a circle whose versine is the deflection.

If the girder be constructed with a camber, the curve of the camber will be flattened when the girder is loaded. The amount of deflection will not be affected by cambering the girder.

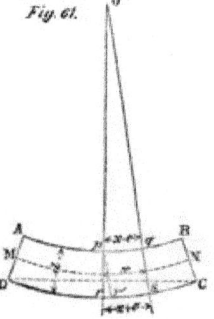

Fig. 61.

Laws of Deflection. Let $ABCD$, fig. 61, represent a girder, originally horizontal, under deflection, having its flanges so proportioned to the strain, that the strain per square inch is uniform throughout the whole of each flange. Then will the flange AB be uniformly contracted throughout its whole length, and CD uniformly extended.

DEFLECTION.

Suppose that on the girder, when undeflected, two parallel vertical lines pr, and qs, be drawn perpendicular to the flanges at a distance x apart, then, when the girder is deflected, their upper extremities p and q will be drawn together to a small extent, which we will call e, so that the distance of p from q will now be $x - e$. In like manner, their lower extremities r and s will be forced apart to an amount e', the distance between them being now $x + e'$. The lines pr and qs being no longer parallel, will, if produced, meet in a point O.

Since the flanges are uniformly contracted or extended throughout their whole length, it is obvious that lines drawn perpendicular to the flanges upon the face of the girder, would, when the girder is deflected, all meet in the same point O; and therefore AB and CD would be arcs of circles, of which O was the centre.

Radius of Curvature. To find the distance Op.

The distance x being assumed very small, pq and rs may be regarded as parallel straight lines.

From point p, draw pr' parallel to qs. Then $r's = pq$, and $rr' = e + e'$.

Since the angle pOq is equal to the angle rpr', the triangle rpr' is similar to the triangle pOq (very nearly).

$$\therefore rr' \text{ or } e + e' : pr :: pq : Op;$$

$$\therefore Op = \frac{pr \cdot pq}{e + e'} = \frac{d \cdot pq}{e + e'},$$

since $pr = d$, the depth of the girder. Whence Op the radius of curvature for the top flange may be found. The neutral axis of the girder, represented by the dotted line MN, does not alter in length.

The radius of curvature of the line MN, or

$$R : x :: d : e + e',$$

by similar triangles,

therefore
$$R = \frac{xd}{e + e'}.$$

When $e = e'$, substitute $2E$ for $e + e'$ in the foregoing expression, and taking $x = 1$,

we have
$$R = \frac{d}{2E} *.$$

E represents the actual alterations of a unit of length, say 1 foot lineal, of either flange under load *in terms of the unit of length, in feet* therefore in this case; d represents the depth of the girder, and R the radius of curvature in the same terms.

To find the Deflection. In fig. 62, let the arc CAD be the neutral axis of the girder when deflected. Let O be the centre, and OA the radius of curvature.

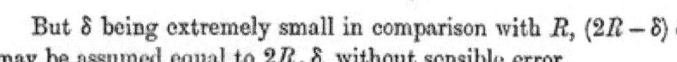
Fig. 62

Complete the circle $ACED$.

Produce AO to E.

By Euclid, Book III. Prop. 35, $AB.BE = BD^2$.

Since $AE = 2R$,

$CD = l$ the span of the girder,

and $AB = \delta$ the deflection,

$$AB.BE = (2R - \delta)\delta = BD^2 = \left(\frac{l}{2}\right)^2.$$

But δ being extremely small in comparison with R, $(2R - \delta)\delta$ may be assumed equal to $2R.\delta$, without sensible error.

$$\therefore \delta = \frac{\left(\frac{l}{2}\right)^2}{2R} = \frac{l^2}{8R}.$$

Substituting for R the expression $\frac{d}{2E}$, we obtain

$$\delta = \frac{l^2 E}{4d} \quad \ldots\ldots\ldots\ldots\ldots\ldots\ldots\ldots\ldots (1).$$

* The method here given of finding the radius of curvature is not absolutely correct, for lines drawn on the side of the girder at right angles to the flanges in its unloaded state will not make the same angle when the girder is deflected, but will make angles with the flanges increasing in acuteness as the deflection and their distance from the centre of the girder increases. Thus the angles ABC, BAD (fig. 61), will be acute angles when the girder is deflected, though right angles when it is undeflected. The consequence is that the true radius of curvature is less, and therefore the deflection greater than our theory, which however is true enough for practical purposes, gives.

Example. A wrought-iron girder 60 feet long, and 10 feet deep, receives a strain of 4 tons per square inch throughout its flanges when loaded; required the deflection.

Since a wrought-iron bar contracts or extends $\frac{\cdot 84}{10000}$ of its length for every ton of strain per square inch of its section, in this case

$$E = 4 \times \frac{\cdot 84}{10000} \text{ feet,}$$

$$l = 60 \text{ feet,}$$

$$d = 10 \text{ feet,}$$

$$\delta = \frac{60 \times 60 \times 4 \times \cdot 84}{4 \times 10 \times 10000} = \cdot 03024 \text{ feet} = \cdot 36 \text{ in. or } \tfrac{3}{8} \text{ in.}$$

$E = CS$, where $C =$ a constant, varying for different materials, and $S =$ the strain in tons per square inch, caused by the load*.

To put equation 1 into a more convenient form, substitute CS for E, then

$$\delta = \frac{l^2 CS}{4d} \quad \dots\dots\dots\dots\dots\dots\dots(2).$$

Irregularly strained girder. When the flanges are not subject to a uniform strain, the foregoing equation is not available, since the value of S is not the same throughout the girder, and consequently the radius of curvature will vary for different parts.

We may, however, assume the value of S to be the same throughout the whole of the flange in each bay, consequently there will be a distinct radius of curvature for each bay.

Let us take the case of a girder in which the value of s is largest for the bays at the centre of the span. Then it is obvious that as S diminishes, the value of E, or $e + e'$, diminishes also (see fig. 61, p. 76), and consequently the radius of curvature R increases in proportion as we approach the abutments.

Let us assume that the girder, when unloaded, lies horizontally, and is constructed and loaded symmetrically on each

---

* If the girder be already partially deflected by a dead load, S will be the additional strain caused by the live load.

side of the centre of the span p (fig. 63), then p will be the lowest point in the girder when loaded, and the centre of curvature for the bay qp will lie in the line po drawn vertically through p. Let o be this point. Join q, o, and produce the line qo indefinitely. Make $qo' = R$ for the bay rq. Join ro' and make $ro'' = r$ for the bay sr. Join s, o''. Through the points q, r and s draw horizontal dotted lines to meet perpendiculars erected from points p, q and r respectively in the points p', q' and r'.

Fig. 63.

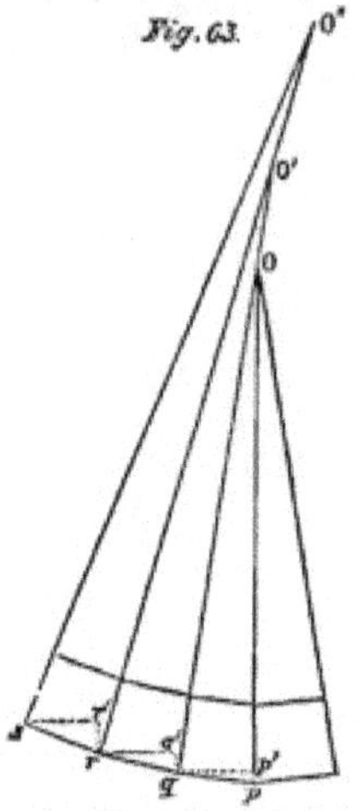

The extremity s of the girder resting on the abutment, it is obvious that the sum of the distances pp', qq' and rr' equals the total deflection of the girder.

If pq, qr and rs be drawn as straight lines, the sine of the angle pqp' is $\dfrac{pp'}{pq}$;

$$\therefore pp' = pq \sin pqp';$$

similarly $\qquad qq' = qr \sin qrq'$,

and $\qquad rr' = rs \sin rsr'$.

Now the angle $pqp' = \angle \dfrac{poq}{2}$;

and ,, ,, $qrq' = \angle poq + \angle \dfrac{qo'r}{2}$,

,, ,, ,, $rsr' = \angle poq + \angle qo'r + \angle \dfrac{ro''s}{2}$;

$\therefore$ the deflection or $\delta = pp' + qq' + rr'$

$$= pq \sin \dfrac{poq}{2} + qr \cdot \sin \left(poq + \dfrac{qo'r}{2} \right)$$

$$+ rs \cdot \sin \left(poq + qo'r + \dfrac{ro''s}{2} \right) \ldots\ldots\ldots(3).$$

In ordinary girders the lengths of the bays are all equal, consequently pq, qr and rs may in such cases be represented by the expression $\frac{l}{N}$, in which l is the length of the girder and N is the number of bays.

Equation 3 may then be simplified to this,
$$\delta = \frac{l}{N}\left\{\sin\frac{poq}{2} + \sin\left(poq + \frac{qo'r}{2}\right) + \sin\left(poq + qo'r + \frac{ro''s}{2}\right)\right\}.$$

The general expression for the deflection of a symmetrical girder of N number of bays is when N is even,
$$\delta = \frac{l}{N}\left\{\sin\frac{\theta^1}{2} + \sin\left(\theta^1 + \frac{\theta^1}{2}\right) + \&c.\ldots\ldots\right.$$
$$\left.\sin\left(\theta^1 + \theta^2 + \theta^3 + \&c\ldots\theta^{\frac{N}{2}-1} + \frac{\theta^{\frac{N}{2}}}{2}\right)\right\}\ldots\ldots(4).$$

When N is uneven,
$$\delta = \frac{l}{N}\left\{\sin\left(\frac{\theta}{2} - \theta\right) + \sin\left(\frac{\theta}{2} + \theta^1 + \theta^2 + \&c\ldots\theta^{\frac{N}{2}-1} + \theta^{\frac{N}{2}}\right)\right\}\ldots\ldots(5),$$
θ being the angle of the centre bay.

To obtain the value of the angles poq, $qo'r$ and $ro''s$, we have only to find the value of R for each bay, which will be given by the equation $R = \frac{d}{2E}$ (see p. 78).

Now the value of any angle θ in degrees may be found from the equation
$$\theta = \frac{\text{arc}}{\text{radius}} \cdot 57°\cdot 29578 ;$$
$$\therefore \angle poq \text{ or } \theta = \frac{pq}{\frac{d}{2E}} \cdot 57°\cdot 29578 = \frac{2E \cdot pq \cdot 57°\cdot 29578}{d}.$$

Substituting $\frac{l}{N}$ for pq and CS for E (see p. 79), we obtain finally for any bay
$$\theta = \frac{2CSl57°\cdot 29578}{dN} \ldots\ldots\ldots\ldots\ldots\ldots(6).$$

In calculating the value of E in plate girders the effect of the web must be taken into account.

For accuracy, therefore, $\frac{1}{6}$th of the area of the cross section of the whole web should be added to the area of the flange (see p. 19).

Deflection of an arched rib. To find the deflection of an arched rib, it will be necessary to know whether the abutments yield sensibly to the thrust when the arch is loaded.

Fig. 64.

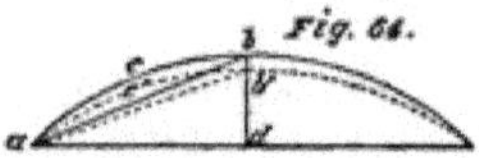

Assuming the abutment to be unyielding, let a, c, b (fig. 64) represent half the rib in its normal position, b being the vertex and a the springing of the arch, of which bd is the rise.

Let the dotted line $ac'b'$ represent the position of the rib when loaded, then bb' is the amount of deflection.

Draw the chords ab, ab'.

Since the distance bb' is very small,

the arc acb : arc $ac'b'$ :: ab : ab', very approximately.

Knowing the strain in compression which the load gives to the rib, the length of the arc $ac'b'$ can be easily found, for

$$\text{arc } ac'b' = \text{arc } acb - (\text{arc } acb \times CS).$$

The half-span ad and the rise db of the arc are given quantities;

$$\therefore ab = \sqrt{ad^2 + db^2} \text{ is known.}$$

But $$ab' = \frac{ac'b' \times ab}{acb} = \frac{(acb - acb \cdot CS)\sqrt{ad^2 + db^2}}{acb}.$$

Now $$db'^2 = ab'^2 - ad^2;$$

$$\therefore db' = \sqrt{\left\{\frac{(acb - acb \cdot CS)\sqrt{ad^2 + db^2}}{acb}\right\}^2 - ad^2}\ldots\ldots(7),$$

and $$bb' \text{ or } \delta = db - db';$$

$$\therefore \delta = db - \sqrt{\left\{\frac{(acb - acb \cdot CS)\sqrt{ad^2 + db^2}}{acb}\right\}^2 - ad^2}\ldots\ldots(8).$$

Example. Required the deflection of a cast-iron arched rib of 20 feet span and 5 feet rise, under load. (See Fig. 65.)

Let the strain on the cross section of the rib average 1 ton per square inch when the bridge is unloaded, and 3 tons when loaded, then the strain caused by the live load or $S = 2$.

Fig. 65.

For cast iron the value of the constant
$$C = \cdot 00018.$$

In this case $\quad db = 5$ feet,

$\quad\quad\quad\quad acb = 11\cdot 6$ feet;

$\quad\quad\quad\quad ad = 10$ feet;

$$\therefore \delta = 5 - \sqrt{\left\{\frac{\{11\cdot 6 - (11\cdot 6 \times 2 \times \cdot 00018)\}\sqrt{100 + 25}}{11\cdot 6}\right\}^2 - 100} \text{ feet;}$$

$\delta = 5 - 4\cdot 99086 = \cdot 00914$ feet $= \cdot 10968$ inches.

Deflection of Bowstring Girder. If the abutments yield, the case becomes analogous to that of the bowstring girder, where the span is lengthened by the stretching of the tie under load.

Let the strong line (Fig. 66) shew the position of the flanges of a bowstring girder when unloaded, and the dotted lines their position when loaded.

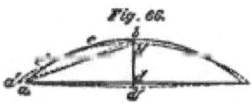

Fig. 66.

Here, as in the previous case,

$\quad\quad\quad$ arc acb : arc $ac'b'$ :: ab : $a'b'$, approximately;

whence $\quad\quad a'b' = \dfrac{a'c'b' \times ab}{acb}$.

But $\quad\quad a'c'b' = acb - acb \cdot CS$. See ante, p. 82.

And $\quad\quad\quad\quad ab = \sqrt{ad^2 + db^2}$,

$$\therefore a'b' = \frac{(acb - acb \cdot CS)\sqrt{ad^2 + db^2}}{acb}, \text{ as before.}$$

Assuming $a'd$ to be a horizontal, and db' to be a vertical line,
$$db' = \sqrt{a'b'^2 - a'd^2}.$$

Now $a'd = ad + ad \cdot CS;$

$$\therefore db' = \sqrt{\left\{\frac{(acb - acb \cdot CS)\sqrt{ad^2 + db^2}}{acb}\right\}^2 - (ad + ad \cdot CS)^2},$$

by substitution.........(9).

And bb' or $\delta = db - db';$

$$\therefore \delta = db - \sqrt{\left\{\frac{(acb - acb \cdot CS)\sqrt{ad^2 + db^2}}{acb}\right\}^2 - (ad + ad \cdot CS)^2}...(10).$$

Deflection of Suspension Bridge. The case of the suspension bridge, though analogous to that of the arch, presents certain peculiarities.

The greatest amount of deflection in a bridge of three spans occurs when the centre span only is loaded.

In consequence of the sliding of the chain on the top of the tower, the chain in the centre opening will become lengthened under load by the amount of extension, not only of itself but also of the back tie.

Wherefore in equation No. 8, p. 82, it will be necessary to substitute for the expression $acb - acb \cdot CS,$

$acb + acb \cdot CS +$ amount of extension of back tie caused by live load.

Strength of a girder how far deducible from its deflection under load. From the deflection of a bridge as actually observed under test load, an idea can be formed of its strength. With the span as the chord of an arc, and the deflection as its versine for data, we can obtain R the radius of curvature, while from the equation

$$R = \frac{d}{2E}, \text{ or } \delta = \frac{l^2 E}{4d} \text{ (p. 78)},$$

the value of E, and the average strain per square inch on the section of the flanges is deducible.

By comparing the deflection of a girder under a moving load with its deflection under a dead load an approximate estimate may be made of the strain to which it is subjected by the moving load. It should, however, be borne in mind that a load moving rapidly on to a bridge deflects it to a greater extent than the same load does when remaining at rest upon the bridge.

The mere stiffness of a girder is no criterion of its strength, for the former depends upon the average sectional area of the flanges, while the latter depends upon the net sectional area at any point. For example, a girder whose flanges were made of the same sectional area from end to end would be stiffer than a girder which had this identical sectional area of flange at the centre of the span and a sectional area diminishing towards the abutments, but it would not be stronger, as the strength would be dependent upon the sectional area at the weakest point—the centre of the span. Again, if one of the plates in the flange of a girder of two or three thicknesses of plate were cut completely through, the deflection of the girder after the injury would not be perceptibly greater than before, though the strength would be seriously diminished.

Mere stiffness alone does not prove the strength of a badly constructed girder.

CHAPTER VIII.

CONTINUOUS GIRDERS.

The term "continuous" is applied to a girder when it is carried without break over two or more spans. The peculiarity of the continuous girder is, that a certain portion of it acts as an ordinary girder, while another part is acting as a cantilever. In the one part the top flange is in compression, and bottom flange in tension, in the other part these conditions are reversed.

Fig. 67 shews the form assumed by two plain girders over adjacent spans when deflected. Assuming that in their unloaded state the flanges of the two girders formed two unbroken parallel

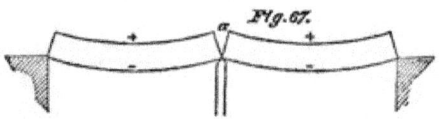

straight lines, and that the ends of the girders at *a* over the pier were in contact for the whole of their depth; upon deflection it is obvious that the extremities of the top flanges of the two girders would part at *a*, and a wedge-shaped gap be formed.

Now let us suppose that before causing the girders to deflect, we had connected together the extremities of their top flanges

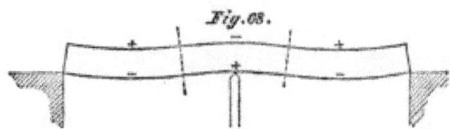

at *a*. The effect of this would be to produce tension in the top flanges of the girders for a certain distance on each side of a, and compression in the bottom flanges. The girders would assume the form shewn in fig. 68, in which the centre of curvature for the flanges in the neighbourhood of the pier is below the girder, while for the other parts the centre of curvature is above the girders. The points where the curve changes from convex to concave, are called points of contrary flexure; at these points there is no strain on the flanges. The parts of the girders in which the top flange is in tension are in fact cantilevers, and the girders of fig. 67, united as in fig. 68, form one "continuous girder."

It is evident that the strains on the flanges will greatly depend upon the position of the point of contrary flexure. This fluctuates with every change in the position of the load, and depends upon the sectional area of the flanges at each point in the girder's length. It is impossible to calculate the exact position of this point in practice, we can, however, define its position within certain limits.

Continuous Girder of two spans. Assuming the girder to be of uniformly proportioned strength, both spans to be alike and similarly loaded, the point of contrary flexure will be at a distance of about $\frac{3}{10}$ ths of the span from the pier, provided the relative level of the three supporting points is similar to that of the three bearing points in the girder in its unloaded state. For instance, if the bearing points of the girder at the extremities and centre are in the same straight line, the supporting points on the pier and abutments must be so also. Or, if one bearing point in the girder is, say 1 inch, higher than the other, its supporting point must be made 1 inch higher than the other points of support.

A corollary may hence be deduced, that the position of the point of contrary flexure in a continuous girder may be shifted at will to a certain extent by raising or depressing any one of the points of support.

The reader may satisfy himself that $\frac{3}{10}$ ths of the span is approximately the true position of the point of contrary flexure in the following manner.

Draw a horizontal line ab (fig. 69) to represent the girder in its unloaded state. Let a be the end of the girder resting on the abutment, and b the bearing point over the pier, the distance ab representing the span.

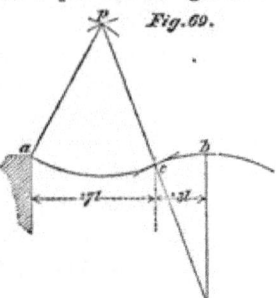

From point b drop a perpendicular bo, making bo of convenient length. With o as a centre, describe the arc bc. With a as a centre, and a distance equal to ob as a radius, describe an arc on the upper side of the line ab.
With o as a centre, and a distance equal to twice ob as a radius, describe an arc cutting the last-mentioned arc in p. With p as a centre, and ap as a radius, describe the arc ac touching the arc bc in c.

A line joining points o and p will cut the arcs bc and ac at their point of contact in c, which is the point of contrary flexure.

If $ab = l$, it will be found by actual measurement that the horizontal distance of the point c from the line ob will be about $\cdot 3l$.

The curved line acb represents the position assumed by an originally horizontal girder under load, for, supposing the girder, as previously stated, to be symmetrically constructed and loaded, it is obvious that the curvature of the girder in each span will be the same, therefore the centre of curvature of the cantilever portion must lie in the centre line ob of the pier, and as it is assumed to be equally strained throughout, the rate of curvature must be the same everywhere, (see Chapter on Deflection). This assumption, however, can never be realized in practice; to do so, it would be necessary to diminish the sectional area of the flanges at c to nothing.

The extra proportional strength of the flanges in the neighbourhood of the point of contrary flexure flattens the curves ac and cb for a short distance on each side of the point c, and would throw c rather further out from the pier. On the other hand, in girders which are constructed to support a moving load, the curvature of the girder portion ac is flatter than that of the cantilever portion, in consequence of the extra metal in the flanges of the girder portion, which is required to enable it to bear the strain which arises when the adjoining span is unloaded and the length of the girder portion increased, as will be seen subsequently: this would throw point c nearer to the pier. For girders that have to bear a dead load only, the points of contrary flexure should be taken as one-third of the span from the pier for safety. This condition of loading gives the greatest possible strain to the flanges at b over the pier, and for girders which have to carry a dead load, such as the wall of a house, no variation in the position of the point of contrary flexure can occur, but when a moving load has to be carried, we must provide for the change caused by the loading of one span only.

Fig. 70 represents a continuous girder of two spans, one only

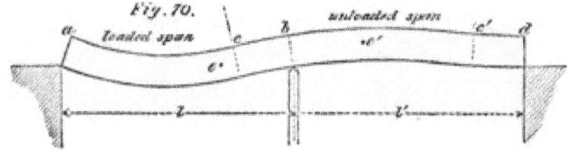

Fig. 70.

of which, that on the left, is loaded. c and c' are assumed positions of the point of contrary flexure; bc and bc' the respective cantilever portions of the girder in the spans l and l'.

The conditions of equilibrium require that

$$\frac{\text{weight of } ac}{2} \times \text{distance } bc + \text{weight of } bc \times \frac{\text{distance } bc}{2} =$$
$$\frac{\text{weight of } dc'}{2} \times \text{distance } bc' + \text{weight of } bc' \times \frac{\text{distance } bc'}{2}.$$

Let w be the unit of the dead load,

,, w' ,, ,, ,, live ,,

Then

$$\frac{ac\,(w+w')}{2} bc + bc\,(w+w') \frac{bc}{2} = \frac{dc' \cdot w}{2} bc' + bc' \cdot w \frac{bc'}{2},$$

or $\qquad \dfrac{(ac+bc)\,(w+w')}{2} bc = \dfrac{(dc'+bc')\,w}{2} bc'$,, ,, ,, but $ac + bc = l$, and $dc' + bc' = l'$,

$$\therefore l(w + w')bc = l'.w.bc' \quad \text{......................} (1).$$

If $l = l'$, $\quad (w + w')bc = wbc \quad \text{...........................} (2)$,

whence $\quad bc : bc' :: w : (w + w')$.

That is to say, when the spans are equal, the lengths of the cantilever portions of each span are inversely proportional to the loads on the spans.

bc', the cantilever portion of the unloaded span, may be any length between $\cdot 3l'$ and l'. It will be the former when $w' = 0$, but what value of w' will give $bc' = l$ can only be determined approximately by very careful calculation. w', however, is generally a known quantity.

Referring to fig. 70, it will be seen that the load on the abutment d is equal to half the weight of the girder portion $c'd$. Now if the point of contrary flexure c' be assumed to coincide with d, there will be no load upon the abutment d. This condition gives the greatest possible length for the cantilever bc, when the spans are unequally loaded.

In calculating the length of the girder portion ac, bc' should always be assumed as considerably under the maximum length l'.

Let e and e' (fig. 70) be the positions of the points of contrary flexure, when the girder abd is fully loaded, then we see that the effect of removing the live load from one span is to cause a considerable strain upon parts of the girder which are unstrained when both spans are fully loaded.

The positions of the points of contrary flexure c and c' will plainly be dependent upon the curvature of the girder, which is regulated by the transverse stiffness. The transverse stiffness again varies directly as the area of metal in the flanges.

In attempting to find approximately the true positions of the points of contrary flexure it will therefore be necessary to assume a certain transverse stiffness of girder, *i.e.* a certain area of metal in the flanges of each bay.

Method of finding position of Point of Contrary Flexure. The method of procedure should be as follows. Assuming both spans to be fully loaded, calculate the strains on the cantilever portions on each side of the pier

CONTINUOUS GIRDERS.

Now assume each span in turn to be loaded, the adjoining span being unloaded.

Take c', the point of contrary flexure for the unloaded span (see fig. 70), as two-thirds of the span l' from b, the pier, and calculate the position of c in the loaded span from equation 2, and the strains on the girder portion ac, and cantilever portion bc'.

Construct a diagram of the girder, shewing the maximum strains on the flanges of each bay, as deduced from the foregoing calculations.

Arrange the plates so as to give at least the required area in each bay.

Take out the actual area of metal in the flanges of each bay, and add thereto one-sixth the area of the web.

By equation 6, p. 81, find the value of θ for each bay, (see pp. 80, 81).

Proceed to calculate the slope of the flange at b, the pier, with the horizontal, in the following manner.

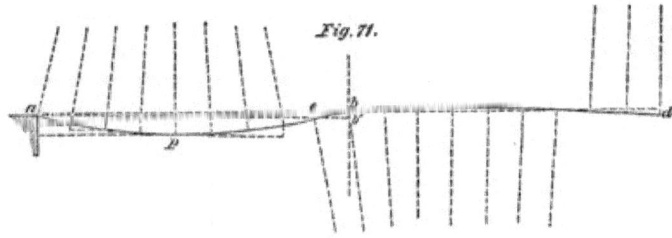

Fig. 71.

Consider the flange at the centre point of the assumed girder portion of the *loaded* span to be horizontal.

Let p (fig. 71) be this point.

Calculate by equation 4, or equation 5, as the case may be, p. 81, the vertical distance of point p below a, the extremity of the girder resting on the abutment, which $= \delta$ in this equation, and also the vertical height of point b above p.

Let ab' be a line drawn parallel to the flange at p, and therefore horizontally, and intersecting a perpendicular line drawn through b in b'.

If the height of a be less than that of b above point p, as in this case, b' will fall below b, but if greater it will be above it.

The value of θ for each bay being known, the angle of inclination of the flange with the horizontal line ab' is also known for each bay.

But the truly horizontal line passes through points a, b and d, consequently the true angle of inclination of the flange at b will be obtained by subtracting the angle bab' from the angle of inclination which it makes with the line ab', when, as in this case, the point b' is below b, or by adding it thereto when it is situated above.

The angle $bab' = \dfrac{bb'}{ab} \dfrac{180°}{\pi}$ in degrees.

Knowing the angle of inclination of the flange at b with the true horizontal line abd, proceed to calculate step by step the angle of inclination of the flange in each bay up to the abutment at d, and thence ascertain the vertical height of the d extremity of the girder above or below the horizontal line.

If it fall above, the girder is safe, but if below, the girder is weak; the remedy for which is, to take the point of contrary flexure c' rather nearer the pier. On the other hand, if the extremity of the girder at d fall much above the horizontal, the point c' may be moved a little further from the pier, and economy effected.

Example. A girder, in two spans of 60 ft., see fig. 1, Plate IX., has to sustain a rolling load three-fourths of a ton per foot run, its depth is 5 feet, and it is divided up into bays of 5 feet, at which distance apart the cross girders are affixed.

Taking the dead load at one quarter ton per foot run, we have $\dfrac{3}{4} + \dfrac{1}{4} = 1$ ton per foot run of girder, or 5 tons per bay.

Following the method laid down on page 90, we commence by ascertaining the strains on the cantilever portions of the girder when both spans are fully loaded.

In fig. 2, Plate IX., are given the strains obtained by taking the point of contrary flexure one-third the span from the pier, that is, at the line dividing bays 8 and 9.

In fig. 1 are given the strains when the left-hand span only is loaded; c' being taken two-thirds of the span from the pier, c falls in the loaded span at the line dividing bays 10 and 11.

In fig. 3 are given the maximum strains collated from figs. 1 and 2.

In fig. 4 are given the arrangement of plates, areas of metal, values of S and θ for each bay of the girder under the condition of loading strains in fig. 1.

Now $\quad\quad \theta = \dfrac{2CSl}{dN} 57°\cdot 29578$, see page 81,

$S =$ strain per sq. inch in tons,

$$C = \dfrac{\cdot 84}{10000}.$$

Here $\quad\quad \dfrac{l}{dN} = 1;$

$$\therefore \theta = \dfrac{2 \times \cdot 84 \times 57°\cdot 29578}{10000} S = \cdot 009626° \, S,$$

whence the value of θ for each bay is quickly obtained.

Assume the lowest point in the girder portion (coinciding with point p, fig. 71) to be in the line dividing bays 5 and 6, which is the centre of the girder portion, and consider this line to be vertical when the girder is deflected.

Let $\theta^1, \theta^2, \ldots \theta^{12}$ be the respective angles for bays 1 to 12 of left-hand span, and $\theta_1, \theta_2, \ldots \theta_{12}$ be the respective angles for bays (1) to (12) of right-hand span. Then the vertical height of the lower left-hand corner of the girder portion above p, or

$$\delta = 5\left\{\sin\dfrac{\theta^5}{2} + \sin\left(\theta^5 + \dfrac{\theta^4}{2}\right) + \sin\left(\theta^5 + \theta^4 + \dfrac{\theta^3}{2}\right)\right.$$
$$\left. + \sin\left(\theta^5 + \theta^4 + \theta^3 + \dfrac{\theta^2}{2}\right) + \sin\left(\theta^5 + \theta^4 + \theta^3 + \theta^2 + \dfrac{\theta^1}{2}\right)\right\}$$
$$= 5\,(\sin\cdot 0151° + \sin\cdot 04405° + \sin\cdot 07146° + \sin\cdot 09633° + \sin\cdot 11148°)$$
$$= 5\,(\cdot 0002635\,\text{ft.} + \cdot 0007689\,\text{ft.} + \cdot 0012472\,\text{ft.} + \cdot 0016812\,\text{ft.} + \cdot 0019457\,\text{ft.})$$
$$= \cdot 0295325 \text{ ft.}$$

On the right-hand side of p, the vertical height of the flange at the extremity of the girder portion, being in the line dividing bays 10 and 11, above p, or

$$\delta = 5 \left\{ \sin \frac{\theta^6}{2} + \sin \left(\theta^6 + \frac{\theta^7}{2} \right) + \sin \left(\theta^6 + \theta^7 + \frac{\theta^8}{2} \right) \right.$$
$$\left. + \sin \left(\theta^6 + \theta^7 + \theta^8 + \frac{\theta^9}{2} \right) + \sin \left(\theta^6 + \theta^7 + \theta^8 + \theta^9 + \frac{\theta^{10}}{2} \right) \right\}$$

$= 5\,(\sin\,·0151° + \sin\,·04367° + \sin\,·07021° + \sin\,·09412° + \sin\,·10792°)$
$= 5\,(·0002635\text{ft.} + ·0007622\text{ft.} + ·0012254\text{ft.} + ·0016426\text{ft.} + ·0018835\text{ft.})$
$ = ·028886\text{ ft.}$

δ for left side of $p = ·0295325$ ft.
$\delta\ \ $„ right $\ \ $„ $\ = ·0288860$ ft.
Difference ... ... $·0006465$ ft.

Value of difference at pier $= ·0006465$ ft. $\times \dfrac{12}{10} = ·0007758$ feet.

For bays 11 and 12 the centre of curvature is below the horizontal line, and consequently the angle of slope of the flange with the horizontal begins to decrease as we go to the right.

The angle made with the vertical by the line dividing bays 10 and 11 = the sum of the angles of the bays between this and the vertical line at the centre of the girder portion

$\phantom{\text{and}}\quad = \theta^6 + \theta^7 + \theta^8 + \theta^9 + \theta^{10} = ·11090°$
and $\quad \theta^{11} + \theta^{12} \phantom{+ \theta^8 + \theta^9 + \theta^{10}}= ·01853°$
$\phantom{\text{and}===}$ Difference $= ·09237°$

= angle with the vertical made by the line dividing bays 12 and (12) at the pier, and angle of flange with the horizontal subject to correction as below.

Now the height of point b (fig. 71) above point e, the lower right-hand extremity of the girder portion,

$$= 5 \left\{ \sin \left(·1109° - \frac{\theta^{11}}{2} \right) + \sin \left(·1109° - \theta^{11} - \frac{\theta^{12}}{2} \right) \right\}$$
$$= 5\,(\sin\,·10826 + \sin\,·09899) = ·018165 \text{ ft.}$$

But ·018165 ft. − ·0007758 ft. (see *ante*) = ·0173892 ft. the distance bb',

and $\angle bab' = \dfrac{·0173892 \text{ ft.} \times 180°}{60 \text{ ft.} \times 3·1416} = ·0166°$,

$$\therefore ·09237° - ·0166° = ·07577°,$$

the true angle of the flange with the horizontal at the pier.

Turning our attention to the unloaded span, we observe that the highest part of the flange will be in the fifth bay from the pier, bay (8), where it will be nearly horizontal. It will make an angle therewith

$$= ·07577° - \left(\theta_{12} + \theta_{11} + \theta_{10} + \theta_9 + \frac{\theta_8}{2}\right) = ·00304°.$$

The flange of bay (7) will slope with an angle of

$$\theta_{12} + \theta_{11} + \theta_{10} + \theta_9 + \theta_8 + \frac{\theta_7}{2} = ·004085°$$

in the contrary direction.

Therefore the highest point of the flange will be in the line dividing bays (7) and (8).

The rise or $\delta = 5 \left\{ \sin\left(·07577 - \dfrac{\theta_{12}}{2}\right) + \sin\left(·07577 - \theta_{12} - \dfrac{\theta_{11}}{2}\right) \right.$

$+ \sin\left(·07577 - \theta_{12} - \theta_{11} - \dfrac{\theta_{10}}{2}\right) + \sin\left(·07577 - \theta_{12} - \theta_{11} - \theta_{10} - \dfrac{\theta_9}{2}\right)$

$\left. + \sin\left(·07577 - \theta_{12} - \theta_{11} - \theta_{10} - \theta_9 - \dfrac{\theta_8}{2}\right) \right\}$

$= 5 \{\sin ·06759° + \sin ·05061° + \sin ·03311° + \sin ·01606° + \sin ·00304\}$

$= 5 (·0011797 \text{ ft.} + ·0008831 \text{ ft.} + ·0005779 \text{ ft.} + ·0005779 \text{ ft.}$

$+ ·0002803 \text{ ft.} + ·0000530 \text{ ft.}) = ·01487 \text{ ft.}$

or about $\frac{3}{16}$ ths of an inch.

The drop from the summit towards d the right-hand abutment or δ

(since $·004085° =$ slope of flange in bay (7) and $·004085 + \dfrac{\theta_7}{2}$

$= ·00654°$, the slope with the vertical of the line dividing bays (7) and (8))

$$= 5 \left\{ \sin \cdot 004085 + \sin \left(\cdot 00654 + \frac{\theta_6}{2} \right) + \sin \left(\cdot 00654 + \theta_6 + \frac{\theta_5}{2} \right) \right.$$
$$+ \sin \left(\cdot 00654 + \theta_6 + \theta_5 - \frac{\theta_4}{4} \right) + \sin \left(\cdot 00654 + \theta_6 + \theta_5 - \theta_4 - \frac{\theta_3}{2} \right)$$
$$+ \sin \left(\cdot 00654 + \theta_6 + \theta_5 - \theta_4 - \theta_3 - \frac{\theta_2}{2} \right)$$
$$\left. + \sin \left(\cdot 00654 + \theta_6 + \theta_5 - \theta_4 - \theta_3 - \theta_2 - \frac{\theta_1}{2} \right) \right\}$$

$$= 5 \, (\sin \cdot 004085° + \sin \cdot 00779° + \sin \cdot 00933° + \sin \cdot 00938°$$
$$+ \sin \cdot 008465° + \sin \cdot 006055° + \sin \cdot 00244°)$$

$$= 5 \, (\cdot 0000713 \text{ ft.} + \cdot 0001359 \text{ ft.} + \cdot 0001628 \text{ ft.} + \cdot 0001637 \text{ ft.}$$
$$+ \cdot 0001477 \text{ ft.} + \cdot 0001057 \text{ ft.} + \cdot 0000426 \text{ ft.})$$
$$= \cdot 00415 \text{ feet.}$$

$\cdot 01487$ ft. $- \cdot 00415$ ft. $= \cdot 01072$ ft. $= \frac{1}{8}$ th of an inch,

the height of the extremity d of the girder above the abutment. But as by hypothesis the end of the girder rests upon the abutment at d, it is obvious that in order to fulfil this condition of things, the angle of the bays between points a and e, fig. 71, should be smaller, or those between e and the point of contrary flexure in the unloaded span, that is to say the angles of the part acting as cantilever, should be larger. Both these effects will result if the points of contrary flexure be taken further from the pier than they are in the present example; thus proving that the length of the girder portion in the loaded span cannot in practice be as long as that we have assumed, and therefore our girder is so far safe, since the maximum strain on the portion over the pier is obtained when both spans are fully loaded, and the maximum strain on the girder portion is less than that we have allowed for.

As the point of contrary flexure in the unloaded span will fall further from the pier than we have supposed, for safety's sake we will assume the unloaded end of the girder as being lifted off the pier. By this means we get the whole of the unloaded span in the condition of a cantilever. By working out the strains produced by this condition of things we find that on the flanges of the third and fourth bays, counting from the pier each way, there will be strains of $52\frac{2}{3}$ and $41\frac{1}{4}$ tons respectively, instead of $36\frac{1}{4}$ and $31\frac{1}{4}$ tons, as given in fig. 3, Plate IX. To meet this it will be necessary to have two thicknesses of $\frac{1}{4}$ inch plate in bays 9 and (9), and to continue the $\frac{5}{16}$ plate through in bays 10 and (10).

The question now arises whether we might not save metal.

By the proportion $bc : bc' :: w : (w+w')$ (p. 90), we get $bc = 15$ft. when $bc' = 60$ ft., that is, when c' coincides with the extremity of the girder at d (fig. 70). It would, however, be manifestly unsafe in fixing the position of the point of contrary flexure in the left-hand span to assume the girder to be lifted off its bearing at d, when the left-hand span only is loaded, for an almost inappreciable error in the fixing or construction of the girder would prevent this from taking place, and the result would be excessive strain on the girder portion of the loaded span arising from increased span. Again, the end of the girder might be lifted off the abutment to an extent that would produce a dangerous jar when the load came on to the unloaded span: for a species of blow would then be given to the abutment by the underside of the girder, destructive of both. If we assume $bc' = 50'$, the utmost we can with safety, then $bc = 12'\ 6''$.

That is to say, the point c would fall in the centre of bay 10, and our girder would be $2'\ 0''$ shorter than what we have assumed.

Very little metal would be saved hereby, and that at increased risk. An error of but $\frac{1}{8}$ of an inch in the relative height of the pier and abutments would only be just covered by our arrangement in Plate IX. As it is, we know c' must fall further away from the pier than we have assumed. The extra metal in the girder portion, by lessening the value of θ for the bays of the girder portion, helps to keep c' nearer the pier than would be the case were the metal in the girder portion reduced to the utmost limit of safety.

We conclude then, that considering all circumstances, it would not be safe to take c further from the pier than we have done.

From the foregoing example the reader will perceive how much of uncertainty and guess-work belongs to the calculations for continuous girders. In actual practice engineers do not think of thus elaborately treating every case of continuous girder that comes before them. They know by experience where to locate the points of contrary flexure so as to keep within safe bounds. In small bridges the labour of calculation would not be repaid by the metal saved in the bridge. In large and important bridges the chances of error are very much diminished, because the proportion of dead weight is much increased, and the amount of variation in the position of the point of contrary flexure is proportional to the difference between w and $w + w'$, as shewn on p. 90.

In our example the variation of the point c amounts to 10 ft. or $\frac{1}{6}$ of the span, and c' is taken as distant $\frac{2}{3}$ of the span from the pier. In calculating for a large bridge where the dead load reaches a high figure, we should be careful not to take c' too far from the pier. We might perhaps take it so far as to get the cantilever portion of the loaded girder as long, or even longer than, one third of the span; and as we know it must be less than this when the adjoining span is unloaded, we know that c' must be assumed at such a distance from the pier as will give it a reduced length.

No universal rule can be given. The safest plan for the student will be to work out carefully a few cases for himself, after the manner of the preceding example, taking different proportions of live and dead load, and thus satisfy himself as to what should be the proportions of cantilever and girder in all cases likely to occur in practice.

Two unequal spans. When the spans are unequal the determination of the position of the points of contrary flexure is still more laborious.

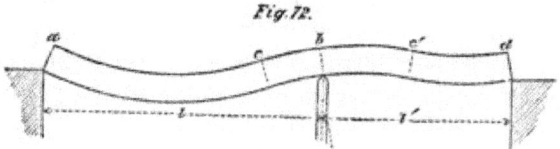
Fig. 72.

Let fig. 72 represent a continuous girder of two unequal spans l and l', having a uniformly proportioned strength.

When both spans are fully loaded, in which case we may assume the unit of load to be alike for both, we find by equation No. 1 (p. 90)

$$l \cdot bc = l' \cdot bc';$$

whence the proportion $bc : b'c :: l' : l$.

In this case the centre of curvature for the cantilever portion will not, as in that of the girder of two equal spans (p. 88), fall in the centre line of the pier. For since the rate of curvature for all parts of the girder is assumed to be alike, it is obvious that the flange at c will make a greater angle of slope with the horizontal than the flange at c', l being the greater span, and as moreover the pier is nearer to c than to c', it is plain that the flange at b must make an angle with the horizontal, and the perpendicular to the flange in which lies the centre of curvature, must fall to the right of the pier, as shewn by the dotted line in fig. 72.

To find point of c. flexure when both spans are loaded. The following is a rough practical method of determining the position of the points of contrary flexure when both spans are fully loaded.

Draw with the same radius three curves touching one another, and forming an undulating line similar to $acbc'd$, fig. 73.

Join the centres of adjoining curves by straight lines cutting the curves at their points of contact in c and c'.

Divide the distance cc' into two parts bc and bc',

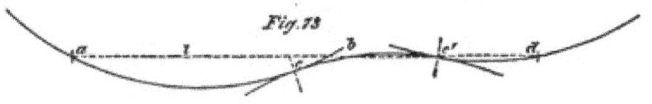
Fig. 73

such that $\qquad bc : bc' :: l' : l.$

Through the point b draw a straight line cutting the two side arcs in a and d, so that

$$ab : bd :: l : l'.$$

This may be done by fixing a needle in the paper at b, and keeping against it the edge of a straight edge, which should be moved about until its edge cuts the arcs at a and d, so as to give the proper proportion between the spans.

By the diagram so obtained the proportion between the cantilever and girder portions of the structure is made visible at a glance.

We have now to determine the positions of the points of contrary flexure when one span only is loaded.

Adopting the notation made use of in the case of fig. 70, we call the point of contrary flexure in the loaded span c, and that in the unloaded span c'.

7—2

Longer span loaded. When the longer span only is loaded the position of c' in the shorter span should be taken nearer to the pier than in the case of equal spans, and the greater the disproportion in the spans the more should c' approach the pier. The object of this precaution is to prevent the end of the unloaded span d being lifted off the abutment, a contingency very likely to occur if there is a deficiency of metal in the girder portion of the longer span, in consequence of the large angle of slope which the flanges would then make with the horizontal at the point c.

The actual position of c' in practice will, however, be nearer to the abutment d than in the case of the equal spans. From this it will be understood that the girder portion of the longer span will be subject to less strain than the other parts of the structure.

Extremely disproportioned spans. If the disproportion between the spans be very great it may be impossible to prevent the unloaded end of the girder from lifting off the abutment.

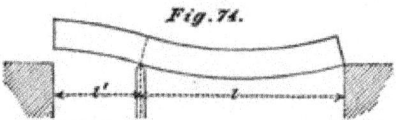
Fig. 74.

Fig. 74 explains this pictorially. Here the structure over the longer span is treated as entirely girder, which occurs when bc has a minimum value $= 0$. Notwithstanding this the slope of the flange at the pier is so great as to lift up the end of the shorter girder from the abutment. By putting a great deal of extra metal in the longer span it might be possible so to flatten the curvature of the girder as to avoid this defect, but as even now we have sufficient metal in the structure to enable the longer span to support itself alone, it is obvious that the proper thing is to span the two openings by unconnected girders.

Shorter span loaded. To find what portion of the shorter span when it alone is loaded should be taken as girder, determine the position which the point of contrary flexure would occupy if the adjoining span were equal in length to the shorter span and unloaded.

The true position will be somewhere between this and its position when both spans are loaded. The greater the disproportion of the two spans the nearer it will approach to this latter point, and of course the longer will be the cantilever portion.

We cannot do more than indicate the law which regulates this variation in position of the point of contrary flexure. The engineer must exercise his own judgment in fixing its position. The limits, as we see, are pretty well known. The only precaution to be observed is not to take the point of contrary flexure too far from the pier, when estimating the strains on the girder portion. Of course it would be possible to find approximately the positions of the points of contrary flexure by calculation in the manner of the last example, but the labour would be doubled in cases where the spans are unequal, and the result would be rendered inaccurate by such slight irregularities in the methods of constructing or fixing the girder, that it would be a waste of time to make the attempt.

Anchoring down the shorter girder. It is sometimes advisable, when the spans are very disproportionate, to obtain the effect of continuity by weighting or fastening down the abutment extremity of the shorter girder. In such cases the shorter span acts entirely as cantilever when the longer span only is loaded.

As the most usual cases of anchoring down occur in bridges having two short side spans and one centre long span, we will indicate the method of determining the amount of keying up necessary to give the required position of the points of contrary flexure in such three-span structures; a precisely analogous method will enable the engineer to do the same for a two-span bridge.

In consequence of the facility afforded by a keying apparatus, it is possible to locate the points of contrary flexure in the longer span at convenient places within limits.

Suppose we desire that they should occur at a distance from the pier $= \dfrac{1}{4}$ of the longer span.

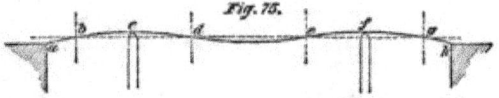
Fig. 75.

Let *abcdefg* (fig. 75) be a representation of the curve assumed by the lower flange of the girder. Then, as the structure will receive nearly its greatest strain everywhere when the centre span only is loaded, we may take for granted that the metal in the flanges has been arranged so as to give an approximately uniform working strain throughout, and consequently that the rate of curvature is the same everywhere.

That being so, it is plain that since the arcs *cd*, *ef* are each of them by construction half the length of the arc *de*, the radii of curvature of all three arcs must be parallel vertical lines; and if distances *cb* and *fg* equal to *cd* and *ef* be marked off on the curve of the flange of the shorter spans, the four points *b*, *d*, *e* and *g* will be in the same straight line.

This line is a certain vertical distance below the bearings on the piers at *c* and *f*, which distance can be ascertained either by calculation or experiment. The formula $\delta = \dfrac{l^2 E}{4d}$ (page 78) will answer for the former method; for the latter, it will be necessary to key up the bridge by guess approximately to its true position, then to load the centre span, and observe by measurement the depression of points *d* and *e*. If the depression given to the points *b* and *g* by the keying up be incorrect, the amount of keying up should be altered until the lines *b*, *d*, *e* and *g* are in one straight line.

As most bridges are built with camber, to simplify matters it is well to set out on the web or upright gussets of the girder when unstrained a horizontal line to be the sole guide for observation. When a bridge of this kind is loaded all over, the curves *ac*, *fh* become slightly flatter as the strain on these parts is somewhat relieved. The effect of this is to shift the points *d* and *e* a little further from the pier.

Multiple spans. We come now to continuous girders of three or more spans.

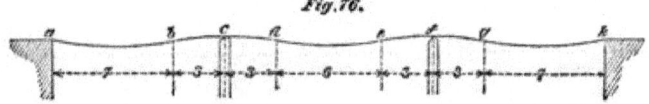

Fig. 76.

Three spans equally loaded. Let the curved line *abcdefgh* (fig. 76) represent a continuous girder of 3 spans and of uniformly proportioned strength, under deflection from a uniformly distributed load.

Let b, d, e and g be points of contrary flexure. Then, if the points of support a, c, f and h are to be in one horizontal line, the span of the centre opening must be to the span of the side openings as 12 is to 10.

For when the centre of curvature of the cantilever portions bd and eg lies in the vertical line passing through the points of support c and f, the length of the cantilevers bc and fg will be $\frac{3}{10}$ of the spans ac and fh (see p. 88). And since the girder is by assumption of uniformly proportioned strength, the radius of curvature of the girder portion de is equal to that of the cantilever portions bd and eg, and consequently the length of de is equal to that of bd or eg; wherefore the whole distance $cf = 4$ times cd.

Now, if $\quad ac = 10$, $bc = 3$, $cd = 3$ and $de = 6$,

$\therefore$ span $ac = 10$, span $cf = 12$ and span $fh = 10$.

If therefore we wish to bridge an opening with a continuous girder of three spans, these proportions of spans will be the most convenient for purposes of calculation and erection. And if we use more than three spans, the intermediate spans should be all alike, and bear the same proportion to the two side spans as does the centre span in fig. 76, viz. 12 to 10.

Spans unequally loaded. It is seldom, however, that multiple-span continuous girders of any size are required to carry only a uniform dead load. In the great majority of cases they are used for railway purposes, where the live load is large in proportion to the dead load. In such cases the most trying position of the live load for the girder will generally be when the two side spans only are loaded, if there are but three spans.

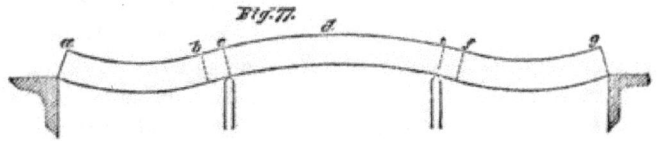

Fig. 77.

The girder $abcdefg$ (fig. 77) is a representation of a continuous girder of three equal spans, of which the two side spans only are loaded. In this case the proportion of live to dead load is assumed to be so great as to throw the point of contrary flexure in the unloaded span so far from the pier as to pass the centre of the span at d. Under these circumstances, both side spans being loaded, there is no contrary flexure in the centre span, but the whole of the top flange is in tension, and the bottom flange in compression.

Fig. 78 represents the same girder under similar conditions of loading, except that the proportion of live to dead load is such

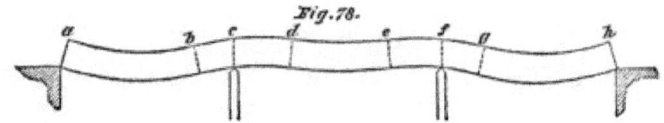

that the points of contrary flexure d, e in the unloaded centre span are at a distance of less than half the span from the piers c and f.

As in the case of the two-span girder, so here it is evident that the position of the points of contrary flexure in the loaded span will depend upon the curvature of the unloaded span.

In order to obtain the position of the point of contrary flexure with approximate correctness, it would be necessary in the first place to assume its position under different conditions of loading,—to arrange the metal in the flanges so as to give sufficient area of metal to take the strain in each bay,—to calculate the amount of bending in each bay, and thence deduce the slope of the flange at the point of contrary flexure in a manner similar to that employed in dealing with the two-span girder, p. 91, and following pages.

If the slope at b (fig. 77), as derived from a calculation of the bending of the portion of the girder bd, agree with that obtained by calculating the bending of the portion ab, the position of the point of contrary flexure which has been assumed is the correct one; but if not, it must be shifted in a direction which will tend to make the slopes so obtained by calculation agree.

The case of fig. 78 is very unusual, and as the variation in the position of the points of contrary flexure in the side spans in such a case is very slight, it is easy to locate them so as to ensure safety without wasting metal. The ordinary case is that of fig. 77.

Diag. 1, Plate X. gives the strains on a continuous girder of three equal spans, when the two side spans only are fully loaded. In this case it will be seen that supposing the point of contrary flexure in the side spans to be at a distance of one-sixth of the span from the piers, the lifting force at the middle of the centre span will be equal to $12\frac{1}{2}$, the strain on the flanges being that which would be caused by a load of $12\frac{1}{2}$, supported at the centre of a girder of the same length as that of the centre span. The span, loading of the girder, and position of the point of contrary flexure in the side span, are taken the same as in Diag. 1, Plate IX. for comparison's sake.

Diag. 2, Plate X. shews the strains on the flanges when the centre span only is loaded, assuming the points of contrary flexure to be one-sixth of the span from the piers, and Diag. 3 gives the strains for the centre span when the whole bridge is loaded, and the points of contrary flexure for the centre span are taken at a distance of one-fourth the span from the piers*. This position gives a strain of only $67\frac{1}{2}$ over the piers, and whereas there is a strain of 60 at these points when the side spans only are loaded, it is clear that the difference between the curvature of the centre span for the condition of a strain on the top flange at the centre of $37\frac{1}{2}$ in tension, and for the condition of $27\frac{1}{2}$ in compression at the same point, would alter the curvature of the side spans more than is implied by the difference of only $7\frac{1}{2}$ in tension over the piers. We therefore judge that the positions of the points of contrary flexure in the centre span should be nearer the centre of the span, or more in accordance with that shewn in fig. 78.

If equality of span be indispensable, economy may be effected by lowering the bearing of the girder on the abutments, so as to throw the points of contrary flexure, b and f (fig. 77), further from the piers.

Reference to Diag. 1, Plate X., will shew that under this condition of loading the centre girder is strained to its maximum

---

* It happens that in the examples given the points of contrary flexure have been assumed to fall at the line dividing two bays, they may however fall at any intermediate point in a bay. When it falls in the centre of a bay there is no strain on the flanges, if the web be doing equal duty in tension and compression. There will be strains in tension and compression on the flanges of the bay according as the point falls nearer the girder side or cantilever side of the bay. Diagrams 4 and 5, Plate X. are examples. In Diagram 5 the point falls $\frac{1}{4}$ of the bay from the girder side of it.

almost throughout its whole length. Now Diag. I, Plate IX. shews that in the case of the two-span girder the unloaded span is, for the greater part of its length, only slightly strained, and consequently its bending is much less than that of the former girder, whence it results that the angle of slope with the horizontal of the flange at the piers is greater in the three-span than in the two-span girder, and therefore the position of c' (fig. 70, p. 89), the point of contrary flexure in the unloaded span, falls nearer to the pier in the three-span girder than in the two-span.

Practical rule for determining the position of the point of contrary flexure in three-span girder. In fixing the position of the point of contrary flexure, the course we recommend is—to determine in the first place the position of the point c (fig. 69), on the assumption that we are dealing with a two-span girder, the unloaded span of which corresponds with the centre span of our three-span girder, in the manner pointed out on pages 89 and 90.

Having done so, to take for the three-span girder the position of point c as somewhat nearer the pier. How much nearer is a matter upon which the engineer must exercise his discretion, as it will be greatly dependent upon the difference between the live and dead loads. If he determines to lower the bearing on the abutment, c need not be taken nearer to the pier.

Girders of more than three spans. When a continuous girder spans four openings, the two centre spans are not subject to so much strain as the centre span in a girder of three spans, for the upward transverse strain upon them is less.

Continuous girders of varying depth. Economy may, in general, be effected by increasing the depth over the piers whereby the area of the flanges at this point is reduced, at the expense, however, of an increase in the amount of web. The greater local stiffness thus produced will affect the position of the points of contrary flexure; its tendency is to throw them further from the piers.

When the load to be carried is entirely dead, or when the proportion of live to dead load is small, the system is undoubtedly attended with considerable economy; but when the proportion of live to dead load is very large, the strains to which the unloaded spans are subjected, in consequence of the greater length of the cantilever portions in the loaded spans, are increased, and much of the saving in the side spans counterbalanced.

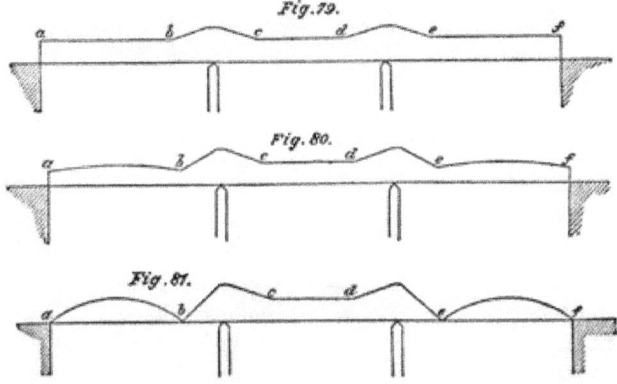

Figs. 79 and 80 represent girders of this kind of three spans; in fig. 79 the girder portions have parallel flanges; in fig. 80 they are hogbacked.

In these examples the position of the points of contrary flexure will fluctuate more or less on either side of the points b, c, d and e, according to the manner in which the girder is loaded.

In the example, fig. 81, the depth of the girder is assumed to be nothing at the points b and e in the side spans, the structure must therefore be hinged at these points. The points of contrary flexure in the side spans are thus fixed under all conditions of loading to b and e. The advantages attending this peculiarity of construction are, that the strains on all parts of the bridge can be calculated with certainty and ease, while the metal, which would be required in the examples of figs. 79 and 80 in the flanges in the neighbourhood of the points b and e to provide for the fluctuation in the positions of the points of contrary flexure, is saved.

The economy and convenience of this system is most conspicuously shewn when circumstances require a large centre span and two short side spans (see fig. 82). As in case of fig. 75 already mentioned, the ends af must be anchored down to the abutments. The points of contrary flexure being fixed, all uncertainty about the strains disappears, and no necessity exists for that extreme nicety in adjusting the bearings of the girder upon which stress

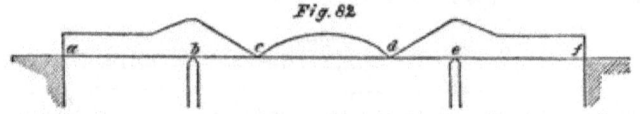

was laid in the case of fig. 75, and which is required in all other kinds of continuous girders.

The structure shewn in fig. 82, properly speaking, does not come under the category of continuous girders; continuity is broken at the points c and d. It must be considered as a compound structure formed of a girder supported by two cantilevers.

Practical remarks. Continuous girders of more than three spans are rarely to be met with. The difficulties of erection and the niceties of construction requisite to ensure the full advantages of continuity are, to many engineers, serious objections to their use. The large amount of contraction and expansion in long girders is another objection.

The attention of the reader is once more called to the importance of careful superintendence in the construction and erection of continuous girders. The chief precaution to be observed is, that the girders be so manufactured that when they are placed *in situ* they shall not be subjected to initial strains except such as the engineer may desire. A simple means of securing this result is to bone a horizontal line on the side of the girder, before it leaves the manufacturer's yard, making a notch with a file in the gussets at the bearings. When the girder rests on its bearings, if it be found that these notches are not in the same straight line, the girder must be packed up, or the bearings lowered so as to make them so, unless it be desired to put an initial strain upon some parts of the girder. The amount of this strain will depend on the deviation of the girder from the original straight line.

As any settlement of the piers or abutments would seriously affect the strains in a continuous girder, it is obvious that very great care must be taken that they are well and substantially built. By means of the horizontal line boned along the face of the girder, settlement of bearings could be observed, and remedied by packing up the girder.

CHAPTER IX.

ROOFING TRUSSES.

THE great variety of forms adopted for the principals of roofs precludes a notice of any but types of the most common.

The most elementary form of principal is that shewn in fig. 83. It consists of two inclined rafters ab and bc, the upper extremities of which abut against each other, while the lower extremities are connected by a tie rod adc, generally cambered at the centre of the span by means of a vertical suspending bar.

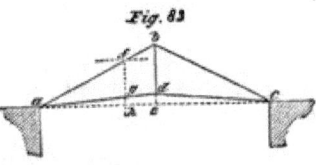

Fig. 83

The strains on a principal of this character may be expeditiously found in the following manner.

Draw through the point of intersection of tie rod and rafter the horizontal line ae.

Erect a perpendicular to the line ae.

With any convenient scale mark off a distance on this perpendicular representing the total load on the abutment, i.e. half the whole load on the principal*.

Through the point so obtained draw a line parallel to ae, intersecting the centre line of the rafter ab in f.

Through f drop the perpendicular fgh to the line ae.

The scale applied to af will give the strain on the rafter: to

---

* By "the whole load on the principal" is meant the load which may be assumed to be concentrated at the apex b in this case. The reaction of the abutment must be assumed to be the share of the load which comes upon it minus the pressure produced upon it by the load carried by the transverse strength of the rafter itself. Thus, supposing the distributed load on the principal abc, Fig. 83, to be 4, the rafter abc (Fig. 83) carries a load of 2 by its own transverse strength, and the point a is kept in equilibrium by three forces, the thrust of the rafter, the pull of the tie-rod, and the vertical reaction of the abutment minus half the load carried by the rafter or $2-1=1$.

ag, the strain on the tie, and to gh, half the strain on the suspending bar.

Or the strains may be found by calculation thus.

As $be : ab : ae : 2de ::$ load on abutment : strain on rafter

:: strain on tie : strain on suspending bar.

Fig. 84 is a diagram of a principal in which the rafters ac and ce are supported at intermediate points, b and d, by struts bf and df. If W be the total load on the principal, there will be a load of $\dfrac{W}{4}$ concentrated at each of the points b, c, and d.

Fig. 84

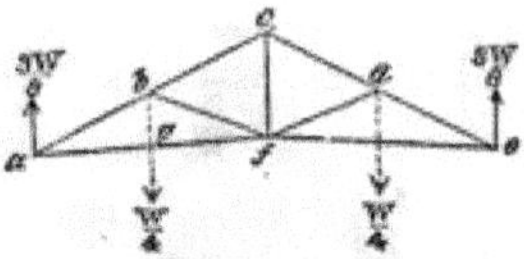

The strains on a truss such as that shewn in fig. 84 might be determined by considering it as one primary, and two secondary trusses, finding the strains on the secondary trusses abf and fdc, and afterwards on the primary truss ace. A similar method may be adopted with the more complicated trusses illustrated in figs. 86, 87, and 88, but we think the simplest method is that indicated on pp. 24 and 25, viz. to find the bending moment at different parts of the truss, and thence deduce the strains on the subsidiary parts.

We propose therefore to adhere to this method as much as possible throughout these investigations.

In the case before us, let us first find the bending moment on the vertical line passing through the point b, and intersecting the tie in the point g. It will be obviously

$$\dfrac{\dfrac{3W}{8} \times \text{horizontal distance between } a \text{ and } g}{bg}$$

$= horizontal$ element of the strain on the tie ag

$= \quad ,, \quad ,, \quad ,, \quad ,, \quad ,, \quad ,,$ rafter ab.

Again, horizontal strain at f

$$= \dfrac{\dfrac{3W}{8}\left(\begin{array}{c}\text{horizontal distance}\\ \text{between } a \text{ and } f\end{array}\right) - \dfrac{W}{4}\left(\begin{array}{c}\text{horizontal distance}\\ \text{between } g \text{ and } f\end{array}\right)}{cf}$$

= horizontal strain on rafter bc.

ROOFING TRUSSES.   111

Now vertical strain on bc + vertical strain on cd

$$-\frac{W}{4} = \text{vertical strain on } cf,$$

and horizontal strain on af − horizontal strain at f
= horizontal strain on bf.

Also for checking purposes vertical strain on bf

$$= \frac{\text{strain on } cf}{2} - \text{vertical strain on } af.$$

Fig. 85 shews another common arrangement of parts for roofs of small span. The bars are marked with the + and − signs, to shew which are in compression, and which are in tension.

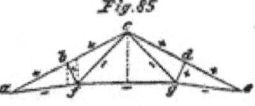

To find the strains on a truss of this shape, it will be necessary to find the horizontal element of the bending moment on the parts at three vertical planes passing through the points b, f, and c, in the manner indicated in the foregoing example.

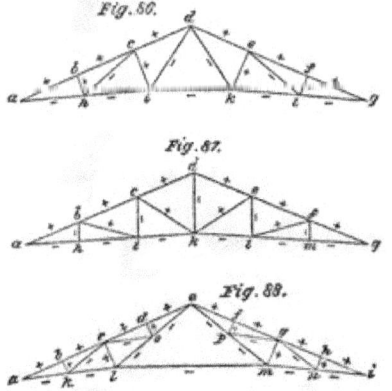

Figs. 86, 87, and 88 are examples of roofs of larger span. The strains may be found by taking the bending moment at vertical sections in fig. 86, through the points b, h, c, i, and d; in fig. 87, through the points b, c, and d; and in fig. 88, through the points b, k, c, and e; for fig. 88, however, the strains on the subsidiary trusses *cdeo* and *efgp* must be obtained separately and added to the strains obtained by the first process.

The strains on the top and bottom members in figs. 86, 87, and 88, being thus ascertained, the strains on the diagonal struts and ties are easily found by methods similar to those adopted for finding the strains on the diagonals of the hogbacked and bowstring girders.

Arched Roofs. Roofs of large span are generally of an arched form.

First System. Fig. 89 represents a plain arched principal intended to act mainly by compression.

If the only load that could come upon the principal were that arising from its own weight, by making the arch of the form of an inverted catenary, a very shallow rib might be used, as the line of pressure would always fall within the flanges; but since all roofs are subject to unequal loading, arising either 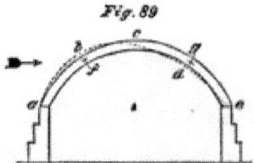 from the pressure of the wind or snow, it does not follow that a catenary will be the best form for our rib. It will also be subject to transverse strain at times, and should therefore have a considerable depth.

To determine the form and scantling of the various parts of the rib.

Having settled upon the amount of rise that it is desirable to have for the crown of the arch, draw the curve of equilibrium for the states of equal and unequal loading through the springing points a and e, and the crown c (see p. 46, and following pages). Draw the dotted line *abcde*, representing the position of the curve of equilibrium when the roof is acted upon by a high wind coming from the left in the direction of the arrow. Suppose the line to fall outside the centre of the rib by a maximum distance bf in the one case, and inside by a distance dg. The most economical form of rib will be that which makes the distances bf and dg equal. Although the rib so obtained may not be graceful, any alteration of form will be at the expense of economy.

The method of finding the strain on the rib at f and g is given on p. 53. The thrust of the rib is taken by the abutments *a* and *e*.

Second System. The resistance to the distortion of the rib may be effected by bracing, as in the ordinary bowstring truss. In roofing trusses the tie is generally very much cambered.

The strains on the top and bottom members of the truss shown in fig. 90, may be obtained by treating it as a girder of varying

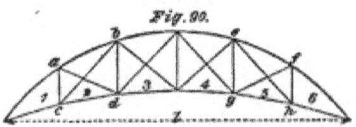

Fig. 90.

depth, the mode we have already adopted when dealing with hogbacked and bowstring girders.

The most convenient way of finding the strain on the diagonals will be to proceed as follows. Ascertain the vertical or shearing force on each *bay* (see pp. 16 and 43) and the strains on the top and bottom members of each particular bay in that condition of loading which gives this shearing force upon it. Resolve the strain of the top and bottom members of the bay into their horizontal and vertical components, the *difference* (in this case) between these *vertical* components of the strains on the flanges will be the total shearing force taken *by them;* this amount must be subtracted from the total shearing force on the bay, *the remainder* will be the shearing or vertical force borne by the diagonals.

If both the diagonals in the bay are struts only or ties only, the whole of this force will be borne by one of them alone. If they are capable of acting both as struts and ties, it will be shared between them. The amount taken by the one or the other will depend mainly upon the sectional areas of metal in the top and bottom members of the bay.

Example. Diagonals, both Ties. Let us by way of example investigate a case in which the diagonals are both ties.

We will assume that when the truss is unequally loaded, it is loaded in a manner similar to that shewn in Diagram 4, Plate VI.

Now the shearing force on bay No. 2 is $24 - 6 = 18$, and the strains on the flanges ab and cd may be found by the method described on pages 25 and 41. Thus:

If l be the span, and $\frac{l}{6}$ = the length of one bay,

$$\text{horizontal strain on } cd = 24 \times \frac{\frac{l}{6}}{ac}$$

$$\text{,, ,, } ab = \frac{24 \times \frac{l}{3} - \left(6 \times \frac{l}{6}\right)}{bd}.$$

The horizontal elements of the strains on the flanges being known, the vertical are easily calculated.

If m = the vertical thrust of the top flange,

and n = the vertical pull of the bottom flange,

$m - n$ = their total shearing effect, since they counteract one another;

and $18 - (m - n)$ is the vertical element of the strain on the diagonal ad.

In this manner the strain on the acting diagonal of each bay may easily be found.

When the left side only of the truss is heavily loaded, the other set of diagonals come into play. Thus, cb and gf will be subject to strain while ad and eh will be unstrained.

Third System. Fig. 91 represents an arched rib with parallel flanges resting on horizontal bearings having one end free to move.

Equally Loaded. When the load acts in a vertical direction only, the strain on the flanges at any point may be easily found by the method described on p. 25 for finding the strain at any part of a girder.

Thus, required the strain on the top and bottom flanges of the rib at point c.

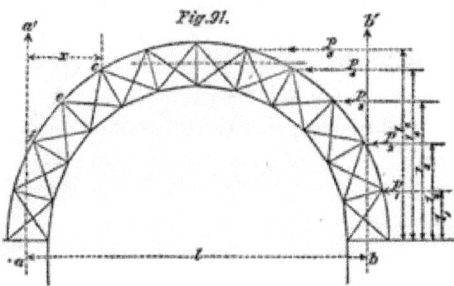

Fig. 91.

We have a force equal to one half the whole load on the rib, or $\frac{W}{2}$, pressing upwards in the direction of the line aa', and acting at a leverage of x, tending to break off the part ac from the remainder of the rib cb. This force is partially counteracted by loads at points e and f resisting the tendency to break at c; therefore the moments of the load at e and f, acting respectively at leverages represented by their horizontal distances from a vertical line passing through point c, must be deducted from the force $\frac{W}{2} \times x$; the remainder divided by the depth of the rib at c will give the strain on the flanges at that point.

The method of ascertaining the strain on the diagonals has just been described (p. 114), and consists in ascertaining the total shearing force to be borne by any particular bay, deducting from that the amount taken by the flanges, and assuming the remainder as borne by one or both diagonals according as they are ties only, or struts also.

Unequally Loaded. The action of the wind upon trusses of this description is very powerful, on account of their great comparative height: it considerably complicates the process of determining the strains upon the various parts. Perhaps the most satisfactory method will be to resolve the force of the wind into its vertical and horizontal components; to treat the rib as subjected to a system of unequal vertical loads, and obtain the strains so arising; then, making another diagram of the rib, to consider it as acted upon by the horizontal forces only. The *sum* of the two systems of strains so obtained will give the actual strain upon each part.

Force of Wind. The force of the wind will act perpendicularly to the curve of the rib; but for the purpose of ascertaining this force it will be sufficient to assume the wind as acting perpendicularly to a tangent to the rib at its junction with the diagonals.

Thus, to find the force of the wind at the point b of the rib (see fig. 92), draw abc tangential to the curve of the rib at b, making the distances ab and bc respectively equal to half the lengths of the bays on each side of the point b. The distance $ac \times$ the distance between the principals is the surface exposed to the wind, and we may take the angle made by abc with the horizontal as that at which the wind acts.

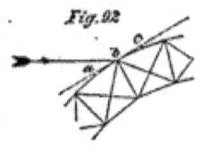
Fig. 92

Pressure of Wind on Inclined Planes. The pressure of wind on inclined surfaces is a subject on which until quite lately we were without any information whatever. Under the auspices of the Aëronautical Society of Great Britain, a few rough experiments have been tried, from which the following table has been compiled as a sufficiently approximate guide to the engineer in calculating the load on a roof from the pressure of wind.

Pressure of wind on plane 1 foot square being taken as 1,

Pressure of wind vertically, not perpendicularly to the plane, on same plane inclined at an angle with the horizontal of—

$15°$ vertical pressure $= \cdot 4$, horizontal pressure $= \cdot 11$;
$20°$ „ „ $= \cdot 5$, „ „ $= \cdot 16$;
$45°$ „ „ $= \cdot 7$, „ „ $= \cdot 7$;
$60°$ „ „ $= \cdot 47$, „ „ $= \cdot 9$.

These experiments were made with smooth planes and low wind pressures, for high pressures and rough surfaces the horizontal pressure should be taken higher than is given in the table.

The maximum pressure of the wind in this country is about 40 lbs. per square foot against an exposed vertical surface. Usually, however, roofs are very much sheltered, and it is only in the case of lofty and exposed buildings that we need assume the pressure so high.

Let p_1, p_2, p_3, p_4 and p_5 (fig. 91) represent the *horizontal element* of the pressures of the wind at the junctions of the bays of the truss, and let l_1, l_2, l_3, l_4 and l_5 be the leverages at which these forces respectively act about the point b, that being taken as the fixed end of the truss and a as the free end.

Then $\dfrac{p_1 l_1 + p_2 l_2 + p_3 l_3 + p_4 l_4 + p_5 l_5}{\text{distance } ab} = Q =$ the vertical pressure on the abutment a, and the lifting force at b.

We have therefore a horizontal force of $p_1 + p_2 + p_3 + p_4 + p_5$ on the bearing at b, and a lifting force of $\dfrac{Q}{2}$ on each of the flanges of the arch at that point. With these data we can, by means of the polygon of forces, find the strains caused by the horizontal force of the wind on the various parts of the structure. When b is the free end and a the fixed end, it will be necessary to commence from the a end of the span, taking as our data the same horizontal force as before, but a vertical pressure of $\dfrac{Q}{2}$ instead of a lifting force of that amount on the extrados and intrados of the arch at a.

It must be remembered to take one system of diagonals only at a time as doing duty, otherwise the problem is insoluble, and to take into account the forces p_1, p_2, p_3, p_4 and p_5 when resolving the forces at the points where they impinge.

The method of estimating the effects of the vertical loads upon the truss, being similar to that adopted in cases of the bowstring and other roofing trusses, will, we trust, need no further explanation.

On Diagram 6 are given the strains on one half of a circular rib due to the horizontal force of the wind calculated by means of the polygon of forces. Being determined by a graphic method, and not by mathematical calculation, they are only approximately correct, but are sufficiently near for practical purposes. The bar aa (Diagram 6) of the truss resting on the abutment is assumed to be held down by its centre to the abutment with a force $= 1610 \times 2 = 3220$, the total lifting force of the wind; it is therefore assumed to be capable of standing a transverse strain of that amount with safety.

The strains caused by the horizontal force of the wind may also be found in the same way as those caused by its vertical effect (see p. 115); for instance, the strain on the bottom flange about the fulcrum b (Diagram 6, Plate X.) may be obtained by multiplying the reactionary upward pressure of the abutment on the right-hand side of the truss by its leverage about point b, and subtracting therefrom the moments about the same point of the forces, 2100 and 900, which tend to turn the truss in a contrary direction (see pp. 4 and 5). The remainder, divided by the depth of the truss, should give the strain on the bottom flange in tension.

CHAPTER X.

ECONOMY IN SUSPENSION BRIDGES.

(Reprinted from "Engineering.")

WHILE mathematicians and engineers of the highest ability have devoted themselves to elucidating the principles which govern the construction of girder bridges of almost every variety of type, but comparatively little of their attention has been bestowed upon the question of economy in bridges on the suspension principle.

This neglect may perhaps be owing to the fact, that, in this country at least, suspension bridges are but rarely constructed, and consequently the element of wonder which attends the construction of bridges of very large span has not yet generally subsided. It is still regarded as so much a feat to throw an iron structure over a clear span of from 600 to 1000 ft., that to criticise its cost would appear ungracious. It is enough that the work has been successfully carried out.

Every one who has given attention to the subject is aware of the important part which the proportion of depth of span plays in the economy of girders of various types, and, as might be expected, it has received due attention at the hands of investigators. But while we are not left in ignorance as to the most economical forms of truss or girder, no one, so far as the writer is aware, has attempted to point out the most economical proportions for the suspension bridge, or the most advantageous method of disposing the materials in its construction.

In view of these facts, it has appeared to the author worth while to make some attempt to supply the deficiency by the elementary investigations presented in this communication, which, wanting as they are in mathematical exactness, are yet, it is believed, sufficiently approximate to be of real value to the practical engineer, and may, it is hoped, lead to further and more elaborate treatment at the hands of the experienced mathematician.

The main questions which it is now proposed to consider are the two following:

1. What proportion should the height of the tower of the suspension bridge above the roadway bear to the span? and

2. What should be the arrangement of parts in the superstructure to obtain the greatest economy?

In endeavouring to answer the first of these questions, let us consider an elementary case.

Let us suppose that we have to support a given load W (fig. 1, Plate XI.), halfway between the sides of a chasm, and are desirous of knowing what is the most economical position for W. All that is necessary is to construct a diagram such as that shewn in fig. 1, Plate XI., in which AB and AC are the suspending bars, whose upper extremities B and C lie at the same level, and whose lower ends meet in A, the point of application of the load.

Let the vertical dotted line AE represent the load W, then the dotted lines ED and EF represent the strains on the bars AB and AC, to which respectively they are drawn parallel; and since the sectional area of each bar is proportional to the strain upon it, by multiplying each bar by its sectional area, a quantity is obtained which represents the amount of material required to carry the load in the given position.

W being equidistant between the points of support B and C, it will be found that the position of greatest economy is attained when the bars AB and AC make an angle of $45°$ with the horizontal, that is to say, when the load is depressed below the points of support to an amount equal to half the horizontal distance between them.

If the point A be raised, the bars AB and AC are shortened, it is true, but the strains ED and EF are more than proportionally increased. On the other hand, if point A be lowered, the strains are diminished, but the bars are more than proportionally lengthened.

The above statement is only true on the assumption that the bars themselves have no weight. If this be taken into account, the most economical position for the point A will be higher than that shewn in fig. I, Plate XI., and the greater the weight of the bars in proportion to the load W, the less will be the depression of the point A below the points of support requisite to obtain the greatest economy.

But few cases, however, occur in practice when it is possible to make use of perpendicular cliffs as points of attachment for the chains of a suspension bridge. It is therefore necessary to take into account the cost of the towers which we have to raise to carry our chains, and the back-ties, in order to arrive at a just estimate of the cost of the whole structure.

For this purpose the diagram, fig. 2, Plate XI. should be constructed, in which (dealing with one side only for simplicity's sake) AB is the suspending bar, BD the back-tie, and BE the tower.

We have now to find that position of the point A which will give $(AB \times$ its sectional area$) + (BD \times$ its sectional area$) + (BE \times$ its sectional area$) - $ a minimum.

Assuming the sectional area of the bars AB and BD, and of the tower BE, to be in the ratio of the strains upon them, this will be attained when the depression of the point A below the top of the tower $B = \cdot 354$ of the distance BC (nearly), that is to say, when the height of the tower is rather more than one-third of the span.

But since in practice, in order to obtain the necessary stability, we require to make the sectional area of the tower much greater in proportion to the strain upon it than is the sectional area of the main chains, it will be obvious that the height of tower just given will be too great to give the most economical practical result, and that the larger the proportion which the sectional area of the tower bears to the strain upon it, the less should be its height in proportion to the span.

Fig. 3, Plate XI. is a diagram from which can be seen at a glance that proportionate height of tower to span which gives the greatest economy for any given sectional area of tower up to 24 times the ratio of section to strain existing in the chains. The vertical distance between the horizontal lines represents the half span. The heights of the tower are represented by the vertical ordinates from the lower horizontal line to the curves, the upper of which terminates the ordinates which give the height of the tower when the load is at the centre of the span, and the lower, the ordinates which shew the most economical height of tower when the load is evenly distributed.

When the load is concentrated at the centre of the span, and there are no towers, that is to say, when the sectional area of the tower $= 0$, the depression of the platform below the points of support should be, as we have already pointed out, equal to half the span, therefore the ordinate at the extreme left of the diagram, with the figure 0 underwritten, is made equal to ·5 of the span. Calling the ratio of section to strain which exists in the chains 1, the height of the tower, when its sectional area bears the same ratio to the strain, is represented by ordinate No. 1. When the ratio of the sectional area of the tower to the strain is twice that in the chains, the height is represented by ordinate No. 2, and so on up to a ratio of 24 times.

From the lower curve we obtain in a similar manner the most economical heights for towers of various characters, when an evenly distributed load has to be carried. These heights are, as might be expected, less than those for towers of similar section, when the load is to be carried at the centre of the span.

To estimate the amount of material in the superstructure of a suspension bridge, such as that shewn in fig. 4, Plate XI., carrying an evenly distributed load and with a varying height of tower, would involve a great amount of labour; it therefore became an object to discover a less laborious method of arriving at a sufficiently approximate estimate of the required quantities. The first step towards the accomplishment of this result was made when it was found that the quantities required to carry a distributed load by the system of fig. 4, Plate XI. were exactly equalled by the quantities required by the system of fig. 5, Plate XI. Finally, it was ascertained by experiment that an identical result was obtained when the whole distributed load was taken as concentrated at two points, each situated at a distance equal to one-fourth of the span from the pier, as shewn in fig. 6, Plate XI.

ECONOMY IN SUSPENSION BRIDGES.

It thus became an easy matter to calculate the quantities required for the superstructure with the different heights of tower.

If AB, fig. 7, Plate XII., be the span, and CD the most economical height of tower of a certain section when the load is concentrated at the centre of the span, the vertical EF will represent the most economical height of tower when the load is concentrated, after the manner shewn in fig. 6, Plate XI., at two points situated at a distance AE from the pier. ·If a series of vertical ordinates representing the most economical height of tower at various positions of the load be raised from the line AB, the point from which each ordinate is raised representing the position in the span of the load to which it corresponds, it will be found that a line drawn through the upper extremities of the ordinates will give a curve similar to that shewn in fig. 7, Plate XII. This curve resembles somewhat the hyperbola. When the point E is situated at a distance from A equal to one-fourth of AB, the ordinate EF will be about five-sevenths of CD, that is to say, for a distributed load the height of the tower should be about five-sevenths of the height when a load is to be carried at the centre of the span only.

The ordinates in fig. 3, Plate XI. are calculated on the assumption that the tower is composed of the same material as the chains, but this is rarely the case in practice; cast iron or masonry being most commonly employed in the tower. In order, therefore, to utilise the diagram, it will be necessary to find the equivalent of the tower in the material of which the chains are composed.

For example: It is required to determine the most economical height of tower for a suspension bridge of which the chains are to be of wrought iron, subject to a maximum strain of six tons per square inch.

Let us estimate for a cast-iron tower. Assuming that the engineer desires that the working-strain on the cast iron shall not exceed three tons per square inch section, and that he estimates the requisite bracing and architectural ornament to absorb twice as much metal as that employed in carrying the load, he will then have one square inch section of metal in his tower for every ton upon it.

Let us assume the cost of cast iron to be in this case one-half that of wrought iron; then we may take the tower as formed of wrought iron, and containing one square inch of metal for every two tons upon it. Now since there is a strain of six tons per square inch on the chains, the ratio of section to strain in the tower is three times that in the chains, therefore ordinate No. 3, fig. 3, Plate XI., will give us the proper height for the tower, which is about eighteen, or rather more than one-sixth of the span.

In a similar manner by obtaining the equivalent of a tower of masonry in wrought iron, the most economical height for a tower of this kind may readily be determined.

In deciding upon the height of the tower, it must not, however, be forgotten, that the greater the height, the greater the expense of erecting the bridge. In order, therefore, to obtain the greatest economy, it will be advisable to find this extra cost in terms of quantity of material. For instance: supposing it is estimated that beyond a certain height from the ground the cost of erection will increase by one shilling a ton for every additional foot of elevation, then if at this height the cost of the material in place be, say £25 per ton, an extra elevation of ten feet would add one-fiftieth to the cost of the bridge, which is the equivalent of an addition of one-fiftieth to the quantities in the structure.

The raising of the height of the tower in this case to the amount of ten feet beyond the given elevation, would only be justified, so far as economy is concerned, when more than one-fiftieth of the quantities was thereby saved.

In fig. 8, Plate XII. is given a series of curves, from the ordinates to which may readily be seen the comparative economy of various proportions of height of tower to span from one-fourth up to one-fourteenth. The figures at the extremity of each curve shew the ratios of the sectional area of the tower to chains, and correspond with the numbers of the ordinates in fig. 3, Plate XI. From this diagram it will be seen that with a tower whose ratio is 3, nearly one-third more material would be required for a bridge with a height of tower equal to one-fourteenth of the span than for a similar bridge whose tower was one-sixth of the span. But when a very expensive character of tower is adopted, the extra cost incurred by keeping the tower low is comparatively small. From this diagram the engineer can tell at a glance what extra expense he will incur by adopting ornamental instead of plain towers.

Fig. 9, Plate XII. shews by means of ordinates to a curve the great economy which can be attained by adopting an inexpensive class of tower. The ordinates represent the minimum total quantities in tower and chains necessitated by the adoption of any ratio of sectional area of tower to sectional area of chains from 0 to 24.

The whole of the foregoing diagrams are based on the assumption that the chains have no weight, they do not, therefore, afford strictly correct information as to the most economical heights of tower under varying circumstances. For when the weight of the structure is taken into account, the load to be carried varies with the height of the tower. The higher the tower, within certain limits, the less the quantities in the superstructure, and consequently the less the load to be carried.

For bridges of very large span, where the proportion of dead load is very large, this variation will be so serious as to affect very materially the question of height of tower. In all cases the most economical height of tower is really greater than that given by the diagrams. In small bridges the error will be inconsiderable, but in large bridges it is quite worthy of being taken into account.

To construct a diagram in which this variation of the load is allowed for, would be a somewhat laborious operation. It has not therefore been thought advisable to go minutely into this question on the present occasion, more especially as a variety of considerations other than that of mere economy often determine the decision of the engineer as to the height of the tower. And whereas the object of the preceding remarks is rather to shew that the proportions of height of tower to span hitherto adopted are much too small, if economy be an object, than to insist upon any particular proportion, it will be enough to point out that if the weight of the superstructure be taken into account, a diagram would result still more condemnatory of the proportions of height of tower to span hitherto generally adopted, than are the diagrams accompanying this article.

Let us now turn our attention to the second part of our subject: "What should be the arrangement of parts in the superstructure to obtain the greatest economy?"

From the point A, fig. 6, Plate XI. erect a perpendicular AB equal to CD, the height of the tower above the platform. Join BC and BE.

Since $BC = AE$ by construction, BE is parallel to AC. So then, if the line AB represent the magnitude of the load concentrated at A, AE or CB will represent the strain on the bar AE, and AC or EB the strain on the bar AC.

Let us now suppose the bar AE, which is subject to tension, to be removed, and its place to be supplied by a strut occupying the position AD. The strain upon this strut and its length are both equal to that of the bar AE, and as the bar AE is common to both arrangements of carrying the load, it follows that if the ratio of strain to sectional area upon the strut be identical with that in the bar AE, the same amount of material will be required whether we make use of the tension bar AE or the strut AD, to carry the load at A.

But if the load be shifted to a point F nearer the pier, the strut FD being shorter than the bar FE while the strain upon both is the same, the strut arrangement will be the more economical of the two. On the other hand, if the load be removed to G further from the pier, the tension bar arrangement will be preferable. It is therefore evident, assuming bars acting in compression to be as effective as bars acting in tension, that the most economical method of carrying a dead load is by means of suspension for that half of the span which lies equally on each side of the centre, and by means of cantilevers for the remaining parts of the span which adjoin the piers.

For a moving load this will not hold good, because, whereas the cantilever parts are not liable to distortion, the same cannot be said of the suspension portion of the structure, to which certain additions are requisite in order to preserve its form under varying conditions of load. It is thus apparent, that for ordinary bridges which have to carry moving loads, it would be necessary to have a preponderance of cantilever, in order to obtain the most economical results.

To determine exactly the proper relative proportions of these two forms of construction in any bridge would be a matter of considerable difficulty. We can, however, conveniently do so under two extreme conditions—the first, when the value of the live load = 0; the second, when the value of the dead load = 0.

The first of these conditions we have just considered; to elucidate the second, let us inquire how the distorting tendency of a moving load can be most economically counteracted.

If we compare the quantities required to carry a given load in the manner shewn in fig. 10, Plate XII., with the systems of figs. 4 and 5, Plate XI., we shall find that the method of fig. 10, Plate XII. consumes about 50 per cent. more material than its rivals. It has, however, this advantage over them, that it is a perfectly rigid system of construction.

If, now, we take the system of fig. 4, Plate XI., and stiffen it by means of two girders of an economical form, in the manner shewn in fig. 11, Plate XIII., we shall find, on taking out the quantities, that they exceed somewhat those of system fig. 10, Plate XII., which may, in fact, be regarded as absolutely the most economical method of carrying by suspension a load entirely moving. We will, therefore, now compare suspension by this system with the cantilever method.

The quantities required to carry the load of 100 by the bars AC and AB, fig. 12, Plate XIII., will be found to be exactly equal to the quantities required to carry the same load by the bar AC and the strut AG, when the distance AG is one-third of the span. The suspension system is the more economical when the distance of the load from the pier exceeds one-third of the span, but the cantilever system when this distance is less than one-third of the span.

We have thus defined the distances of one-fourth and one-third of the span from the pier to be the limits between which it will be necessary to fix the point where the character of the structure should change from that of suspension to cantilever, if economy is to be a supreme consideration.

Now, if we compare together the systems of fig. 10, Plate XII. and fig. 11, Plate XIII., when the load to be carried consists partly of dead and partly of live load, we shall find the latter system (fig. 11, Plate XIII.) to be more economical than the former in proportion to the quantity of dead load which has to be carried. For whereas by the system of fig. 10, Plate XII. the quantities required will be the same, whether the load be live or dead, in the system of fig. 11, Plate XIII., when the live load $= 0$, the quantities in the stiffening trusses become $= 0$, and the bridge takes the character of fig. 4, Plate XI. Hence it is evident that, in ordinary suspension bridges, where the dead load forms a serious item in the load, the system of fig. 11, Plate XIII., will be preferable to that of fig. 10, Plate XII.

From the foregoing facts we deduce the principle, that when the proportion of dead load is excessive, the point where the suspension part of the structure ends and the cantilever portion begins, should approach the limit nearest the pier. On the other hand, when the live load preponderates, it should lie nearer the limit furthest from the pier.

In these calculations the effect of the weight of the structure itself has been completely left out of account. It will, however, modify our conclusions to a certain extent, and in the direction of further increasing the proportions of cantilever.

If we turn to fig. 6, Plate XI., and once more compare the suspension with the cantilever system, we shall see that since the tension bar AE occupies a position nearer the centre of the spans than the strut AD, the weight of the structure is more advantageously situated for economy in the cantilever than in the suspension part. Hence the truly economic position of the point of change from suspension to cantilever method would be at a distance, more or less, exceeding one-fourth of the span from the pier, as the weight of the bars bore a large or small proportion to the load carried.

Thus much for the economical part of the subject. Let us now devote a short time to the consideration of the practical part.

However economical any arrangement of parts may be, its adoption must be barred if it involve danger to the safety of the bridge. It probably has been with the idea chiefly of keeping down the amount of oscillation to which a suspension bridge is liable, that the curve of the chains in many existing suspension bridges has been made so excessively flat. And if no means of stiffening the platform is to be adopted, the arrangement is so far calculated to attain the desired object. But this is a very expensive way of accomplishing in a very imperfect manner a result which can be completely effected by the use of stiffening girders. The transverse strain to which these are subjected is always the same with certain proportions of live to dead load, and is entirely unaffected by the rate of curvature of the chain. There can, therefore, be no objection on this score in a bridge provided with stiffening girders to proportionally high towers.

The higher the towers the less the deflection of the bridge under load, and the less the alteration of form caused by changes of temperature. This, to some engineers, will appear an important consideration.

Let us now consider the behaviour of structure, such as that shewn in fig. 13, Plate XIII., compounded of a suspension portion AB between two cantilever portions AC and BD, under a rolling load.

The load travelling on to the bridge from left to right, a bending will take place at each of the points E, F, and G, in the platform. At these points it should therefore be hinged. The alteration in the form of the platform thus caused will, however, be slight. In a bridge of 1600 feet span, where the proportion of dead load would be large, it was found by calculation that the effect of an ordinary train running on to the bridge as far as point E, would be to depress the platform at that point to an amount of about one inch. We may therefore take it that the distortion of the platform produced by unequal loading would be so insignificant as to be unworthy of consideration.

The alteration in the form of the platform caused by changes of temperature will, however, be considerable. The expansion of the metal will cause the extremities E and G of the two cantilevers to approach one another, and consequently the platform at F to drop. In order to prevent cross strains, it will be advisable to make the hinge joints at E and G such as to admit of a certain amount of end motion in the platform. This can be done in a variety of ways.

In order to avoid the unsightly appearance which would be produced by a reverse camber in the platform, it is recommended that the platform should be laid so as to have a slight gradient each way towards the centre of the span, as shewn by the straight lines H, E, F, F, G, I, fig. 14, Plate XIII., when the temperature is at its lowest. When the temperature is at its highest, the point F would drop to F^n, a point not below the level of the hinges E and G.

By this arrangement, all unsightliness can be avoided. The consequence of these alterations in form of platform would amount simply to this, that the gradients on the bridge would vary with the season of the year, a peculiarity in no way affecting the safety of the structure. Nor would the fact that the platform is capable of end motion at the points E and G affect its lateral stiffness, provided it be treated as a continuous girder, of which HE and GI are the cantilever portions and EG the girder portion, E and G being the points of contrary flexure.

The erection of such a bridge as that shewn in fig. 13, Plate XIII., could be conveniently effected by building out from the piers as far as the extremities of the cantilevers, and erecting the suspension portion in the usual manner.

We have not here entered into the question of the most economical height of tower for a bridge of single span: to do so would unnecessarily lengthen this article; we may, however, consider that the proper height of tower for a bridge of this kind should be much the same as for a bridge of several spans, inasmuch as the quantities in the back tie of the former would be not very different from those in the half span of the latter description of bridge. When the towers of a bridge of single span are backed by high ground, of course it would be advisable to adopt a loftier tower than would be proper if no such natural feature characterised the site.

The result of the investigations briefly summarised in the foregoing remarks has been to convince the author that considerable unnecessary expense has been incurred in bridging large spans, through the adoption of towers deficient in height, and an arrangement of parts in the superstructure not of the most economical description. In confirmation of these opinions he would state that he has made numerous careful and detailed estimates for bridges of various spans on the system of fig. 13, Plate XIII., and the result has fully justified his anticipations.

INDEX.

ABUTMENTS of an arch, yielding of, 83
Alternative methods of finding strains, 10
Anchoring down continuous girder, 101
Arch, 61; thrust at crown of, ib.; methods of resisting distortion, 62; strain on spandrils of, ib.; stability of pier of, 68; effect of horizontal girder in, ib.; transverse strain on piers of, ib.
—— of masonry, curve of equilibrium for, 61
Arched bridges of multiple span, 67
—— viaduct of wrought iron, 70
Arm, 1
Axis, neutral, 18

Back tie of suspension bridge, 73
Beams, wooden, 18
—— of irregular section, 21
Bowstring girder, 45; curve of equilibrium, ib.; method of drawing the curve, 46
—— regularly loaded, 46; irregularly, 48; method of resisting distortion of, 51; load resting on the top, 60

Cantilevers, 17
Continuous girders, 86; of two spans, 87; of two unequal spans, 98; example, 92; method of finding position of point of contrary flexure when both spans are loaded, 99; when longer span loaded, 100; extremely disproportioned span, ib.; shorter span loaded, ib.; anchoring down the shorter girder, 101; multiple spans, 102; three-span girder, 106; girders of more than three spans,

ib.; of varying depth, ib.; practical remarks on, 108
Curve of equilibrium, 45; regular loading, 46; with a given versine, 47; irregular loading, bays equal, 48; with bays unequal, 49

Decomposition of forces, 8
Definition of deflection, 76
Deflection, laws of, 76; of arched rib, 82; of bowstring girder, 83; of continuous girders, 92; of suspension bridge, 84; strength of a girder, how far deducible from, ib.; of an irregularly strained girder, 79; to find the, 76; definition, ib.
Diagonals, effect of elasticity upon, 56; resisting extension and compression, 60

Economy in suspension bridges, 119; proper proportion of height of tower to span, ib.; method of estimating proper height for towers of various kinds, 123, economical arrangement of parts in superstructure of suspension bridge, 126; suspension and cantilever system compared, for dead loads, ib.; for live loads, 127; behaviour of a compound cantilever and suspension bridge under loads and changes of temperature, 128; summary of results, 130
Effect of horizontal girder on arch, 68
Elasticity, effect of, in distributing load among diagonals, 56
Equilibrium, curve of, 45
Evenly distributed load, 17; on cantilever, ib.; on parallel flanged girder, ib.

Forces, resolution or decomposition of, 8; acting in different planes, 11
Fulcrum, 1

Girders with parallel flanges, 12; load at centre of span, ib.; load evenly distributed, 13; irregularly loaded, 16; bays equal, 33; bays unequal, 36; load symmetrical, bays unequal, 35
—— lattice, 37
—— warren, 38
—— hogbacked, 41
—— of irregular section, 21
—— ordinary wrought-iron, ib.
—— bowstring, 45
—— continuous: see *Continuous Girders*

Hogbacked girders, 41; compared with parallel flanged girder, 43
Horizontal girder to arching, 62; effect of on piers in arched viaducts, 68

Irregular loading of arches, 63; of bowstring girders, 48; of parallel flanged girders, 16; of hogbacked girders, 41; of suspension bridge, 73; of girders, 89

Lattice girders, 37
Lever, 1
Lever, bent, 3
Leverage, ib.
Line of greatest resistance, 51
Load at centre of span, 12

Masonry, arch of, 61
Method of finding position of point of contrary flexure in continuous girders, 90
Methods of resisting distortion in the arch, 62; in the bowstring girder, 51
Moment, 2
Multiple span, arched bridges of, 67

Multiple span, continuous girders of, 102

Neutral axis, 18

Ordinary wrought-iron girders, 21; strains on the flanges, 22; strains on the web, 24; shearing strain on each bay, 27

Parallelogram of forces, 6
Polygon of forces, ib.
Practical remarks on continuous girders, 108

Radius of curvature of deflected beams, 77
Resolution of forces, 8
Resultant, 6

Shearing force, 16; formulas for, 30
Spandrils of an arch, 62
—— filling, sundry forms of, 66
Stability of piers, 68
Strain on flanges at centre of span, 13; at any point of the flange, 14; on the web, 16, 24; on the flanges of hogbacked girders, 41; on diagonals, ib.
Suspension bridges, 72; strain at centre of, 73; strain at any portion of chain, ib.; methods of stiffening, 74

Transverse strain on piers of arched viaducts, 68
Triangle of forces, 6
Truss, warren, 38
Trusses, 37

Vertical, or shearing force, 16
Viaduct, arched, of masonry, 67; of wrought-iron, 70

Warren truss, 38; strains on diagonals of, 40
Web in iron girders, 19
Wooden beams, 18

CAMBRIDGE: PRINTED BY C. J. CLAY, M.A. AT THE UNIVERSITY PRESS.

Plate.1

Dia. 1.

Dia. 2.

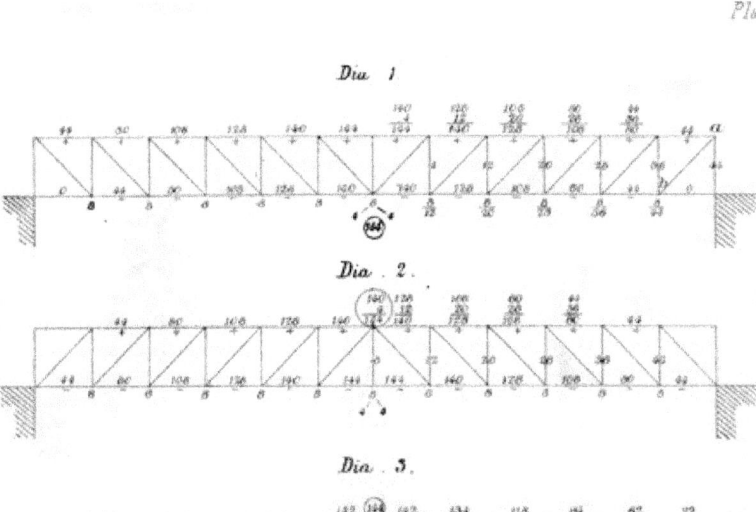

Dia. 3.

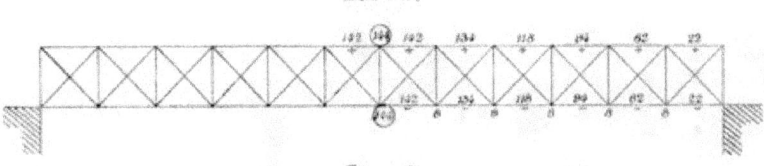

Dia. 4.

Dia. 5.

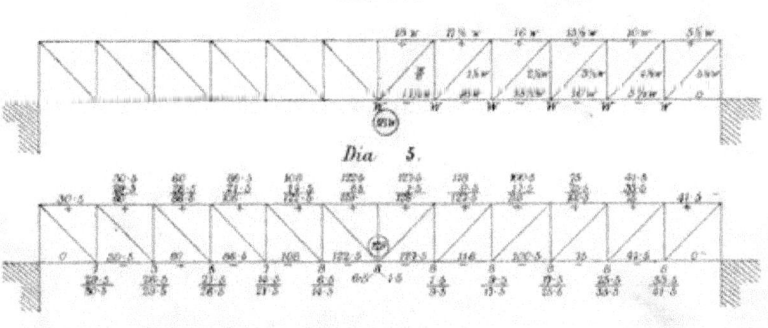

Dia. 6.

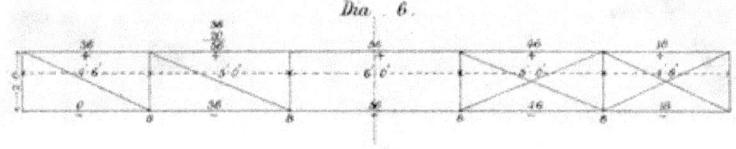

Dia. 7.

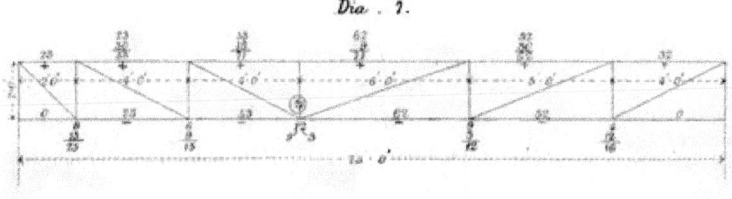

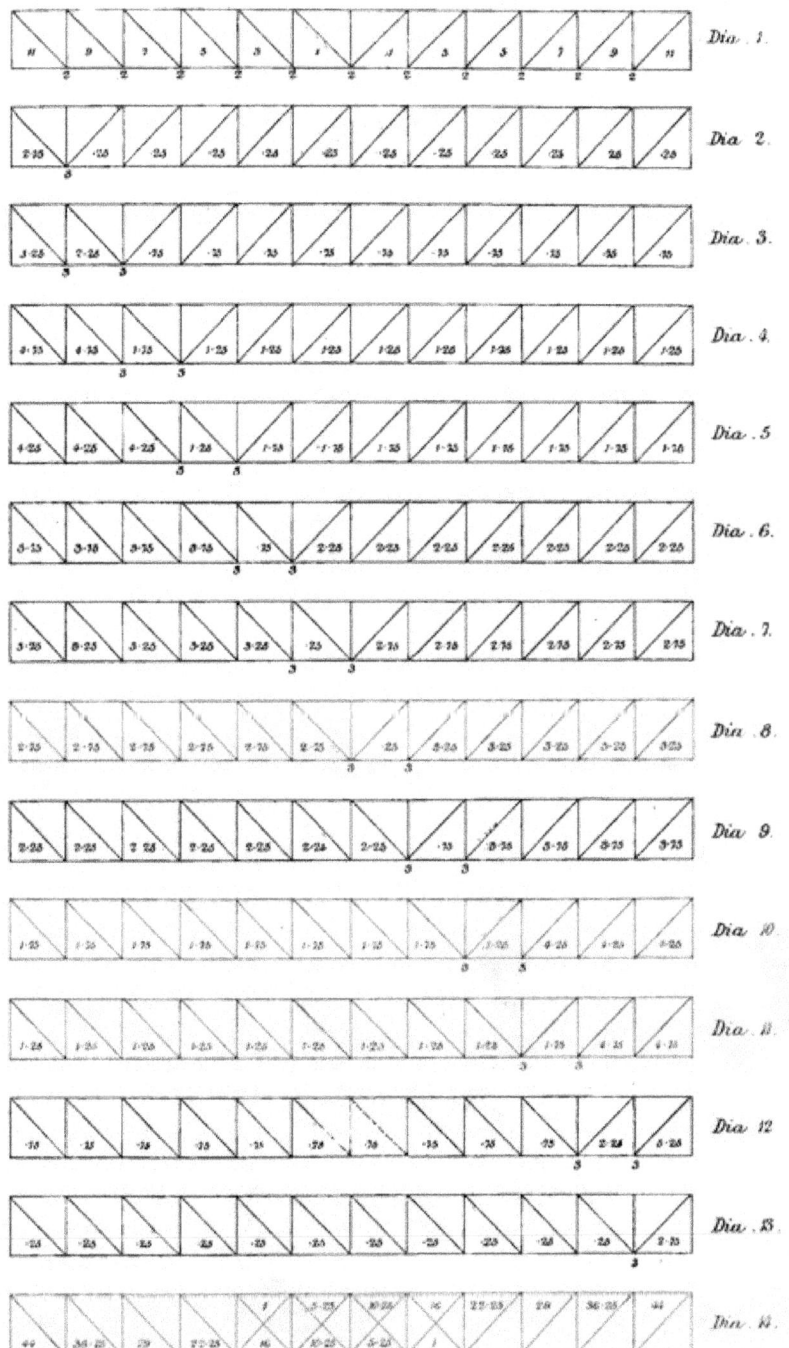

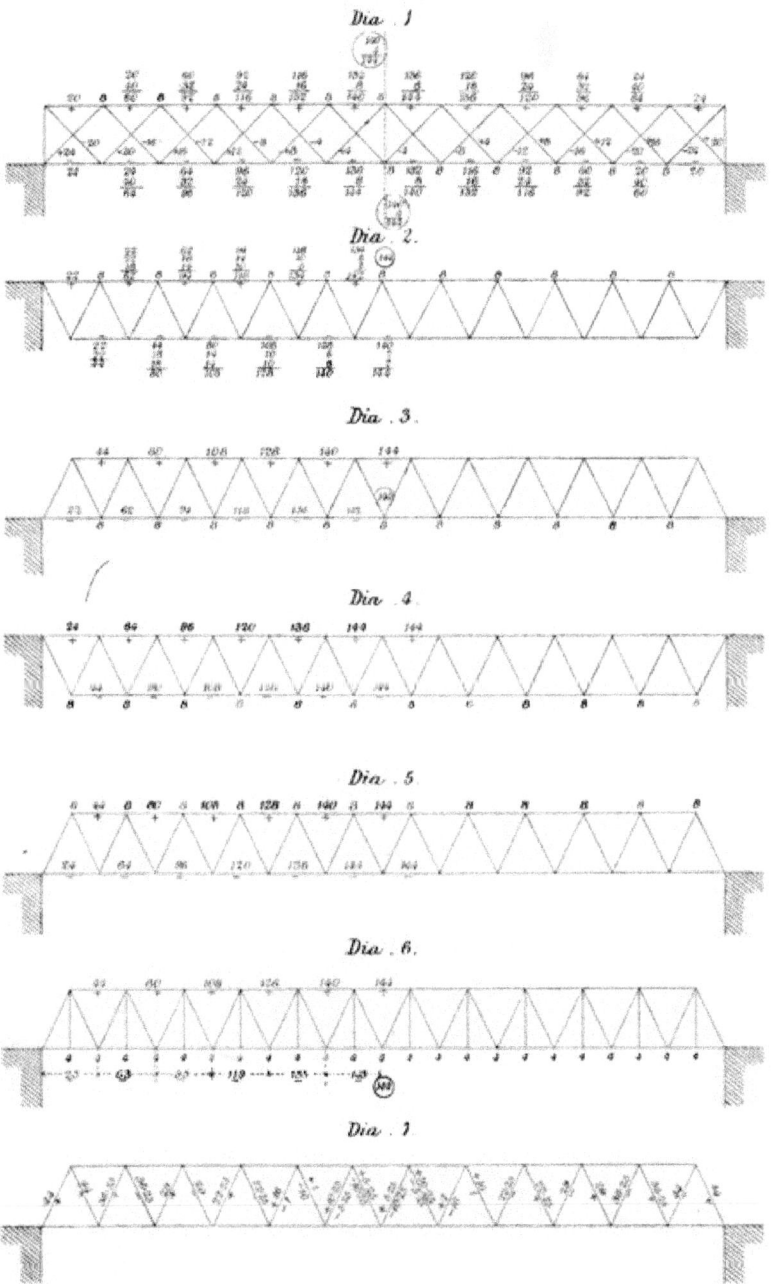

Plate III.

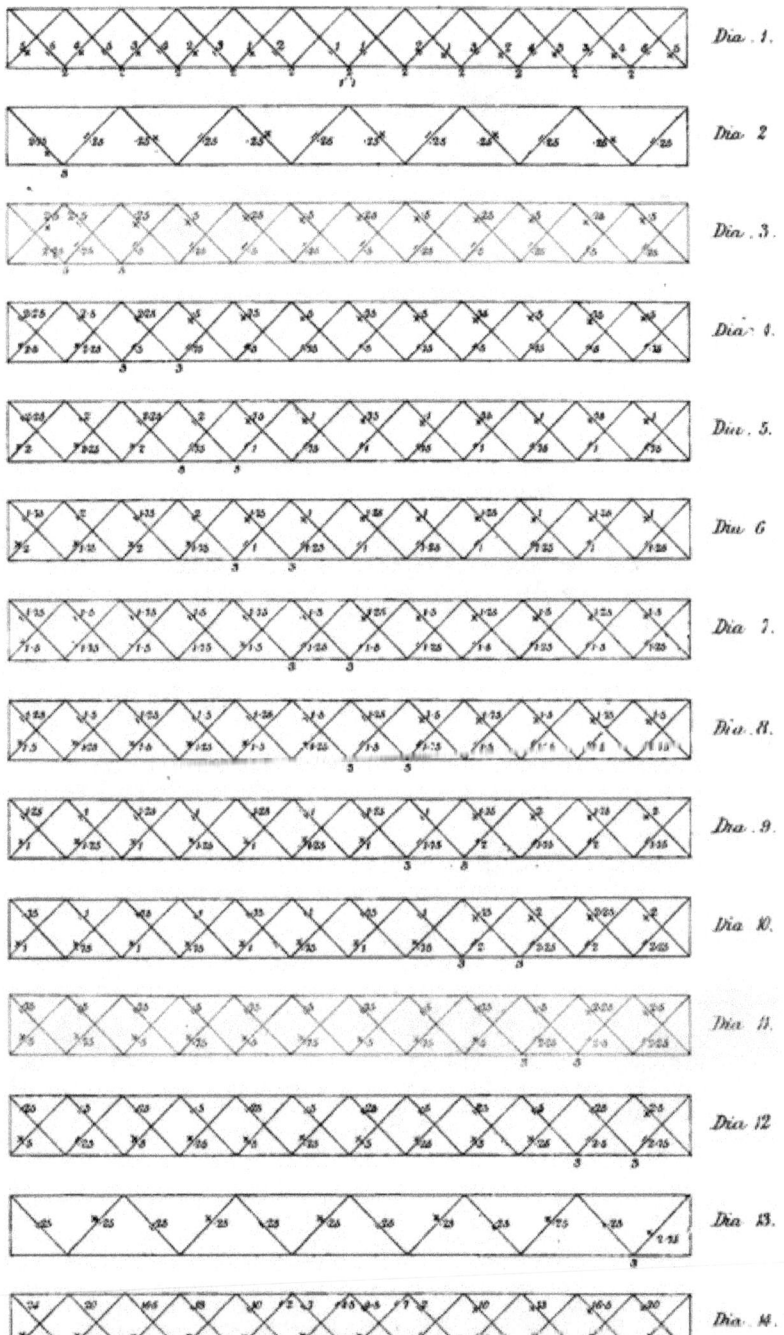

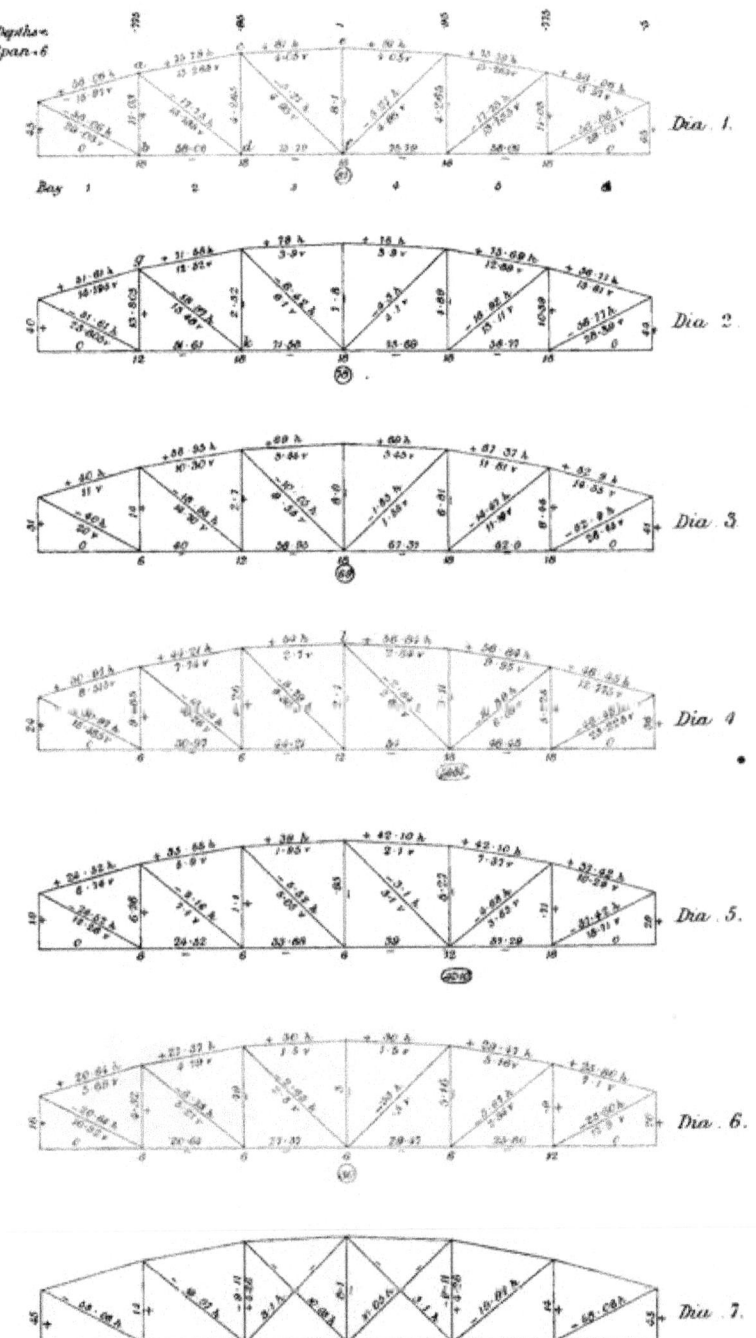

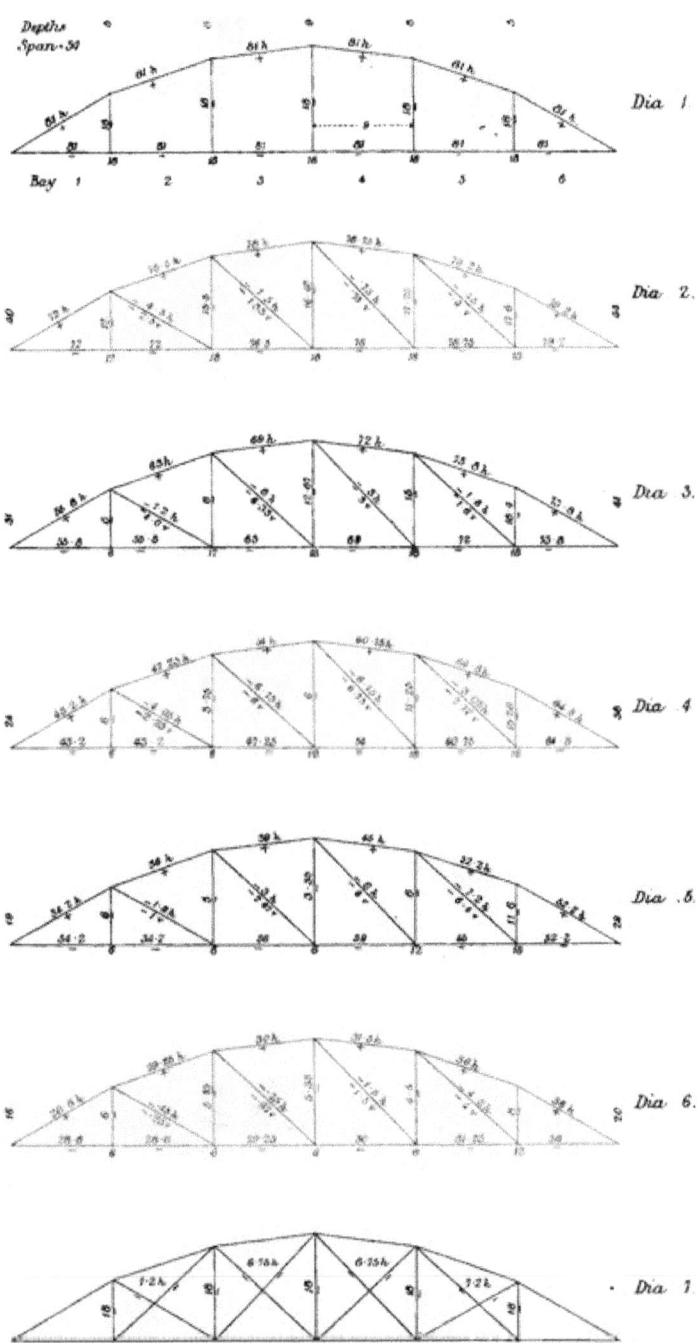

Plate VII

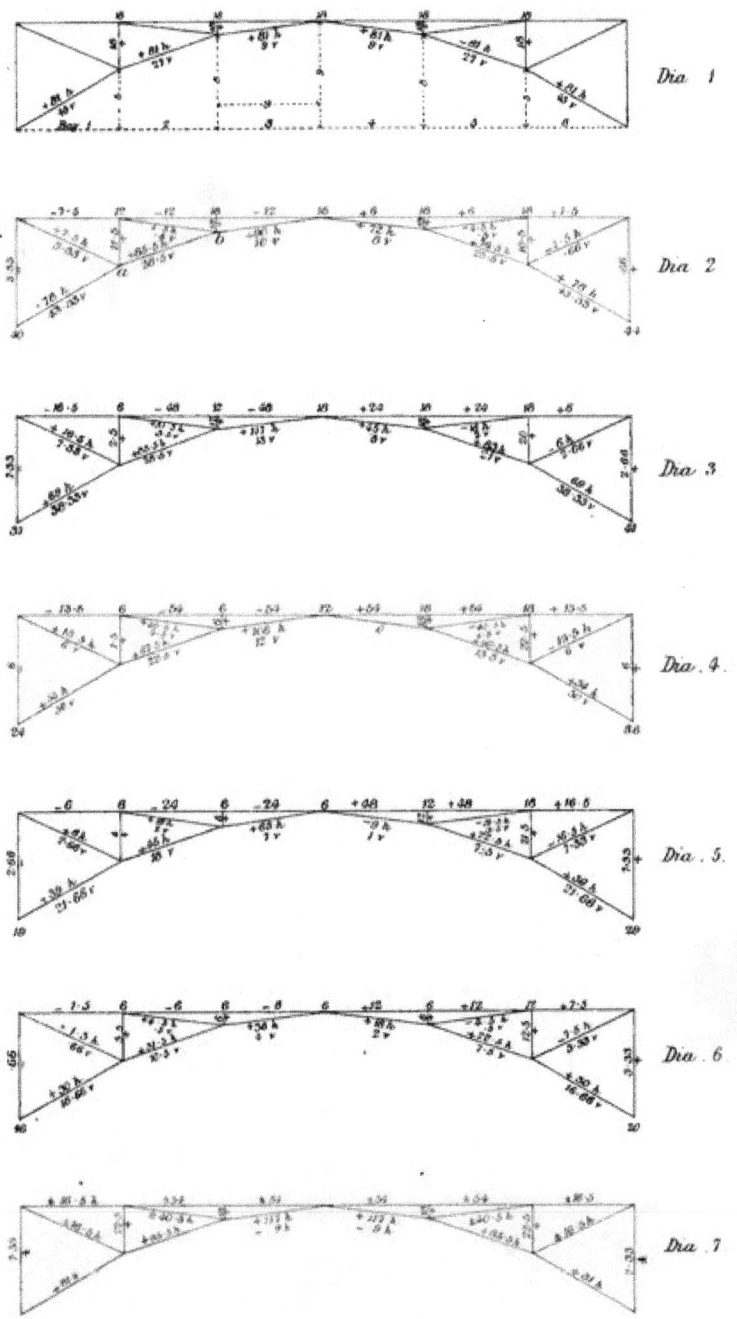

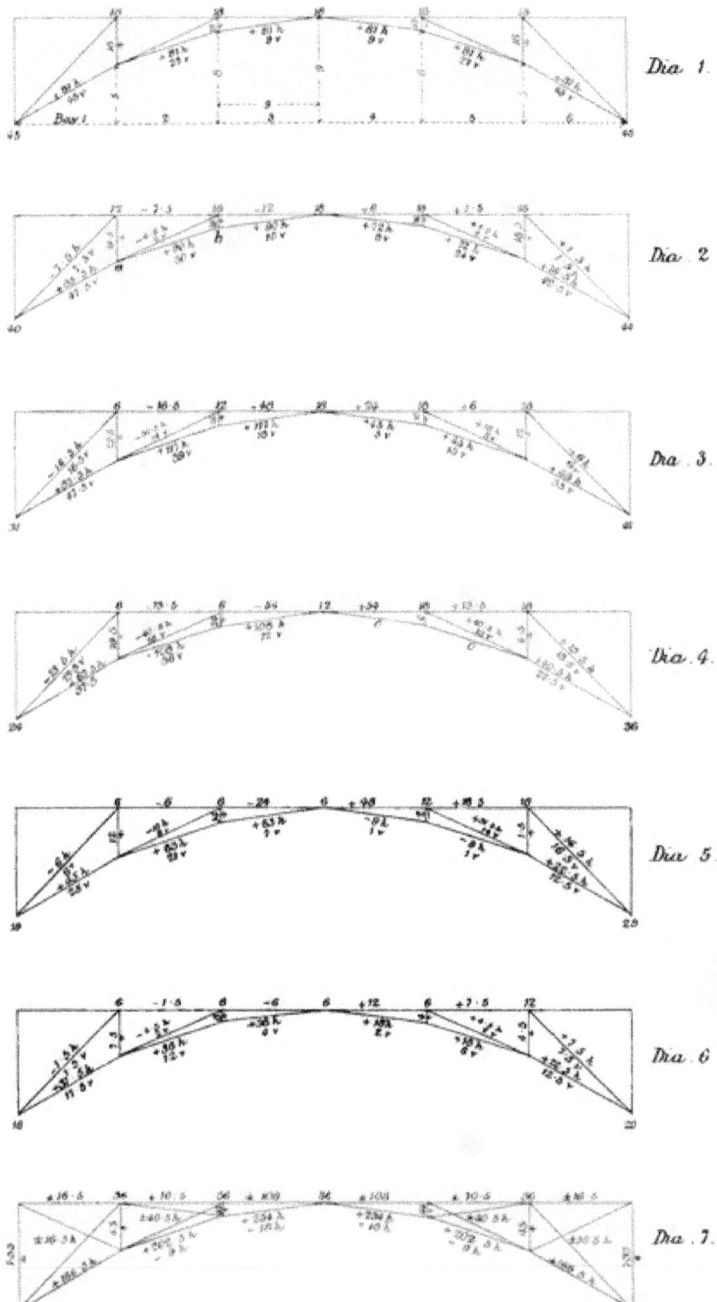

Plate IX

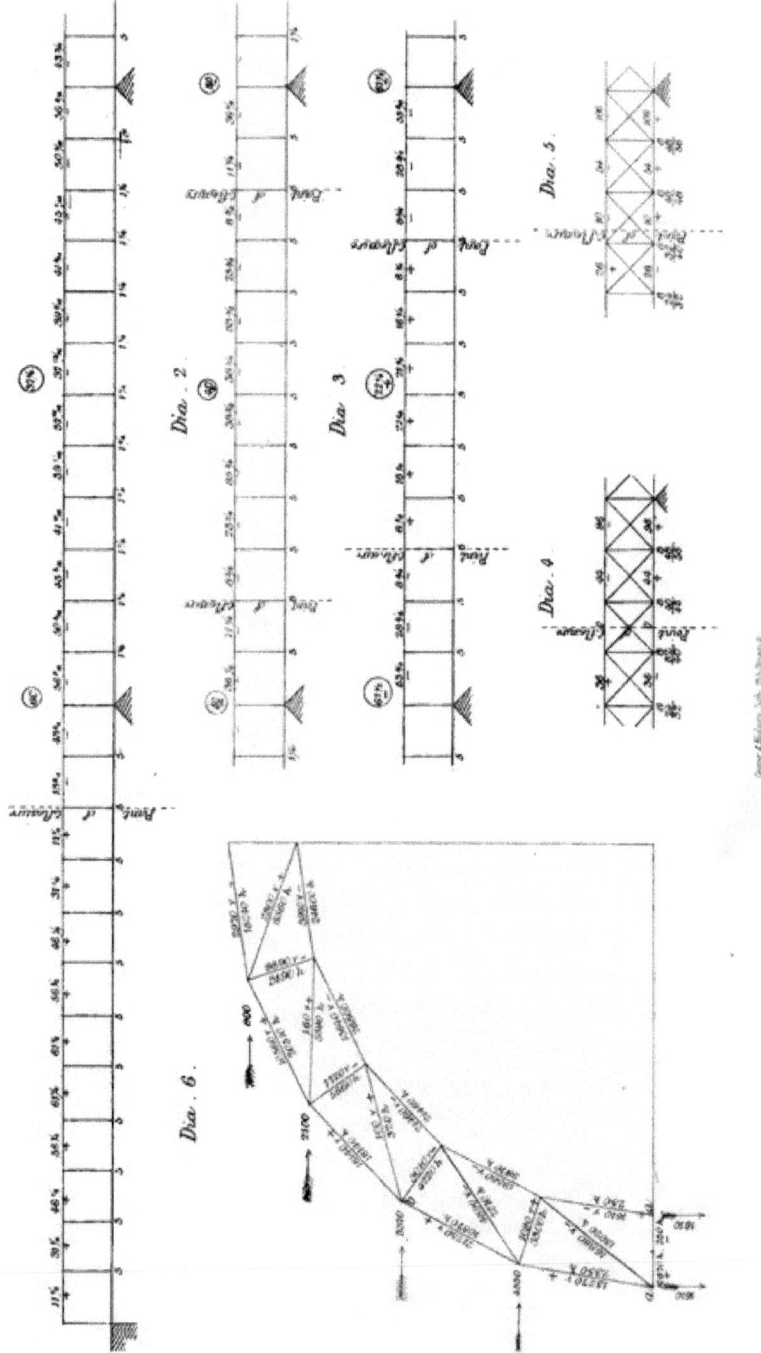

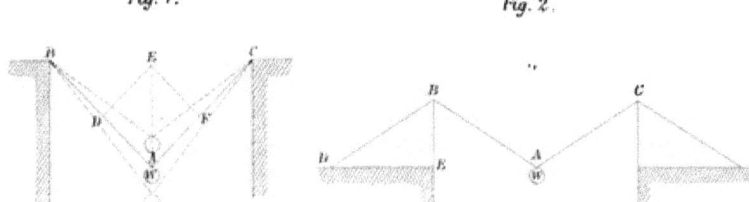

Fig. 1.   Fig. 2.

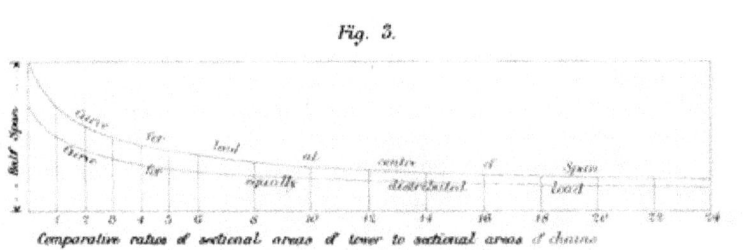

Fig. 3.

Comparative ratios of sectional areas of tower to sectional areas of chains

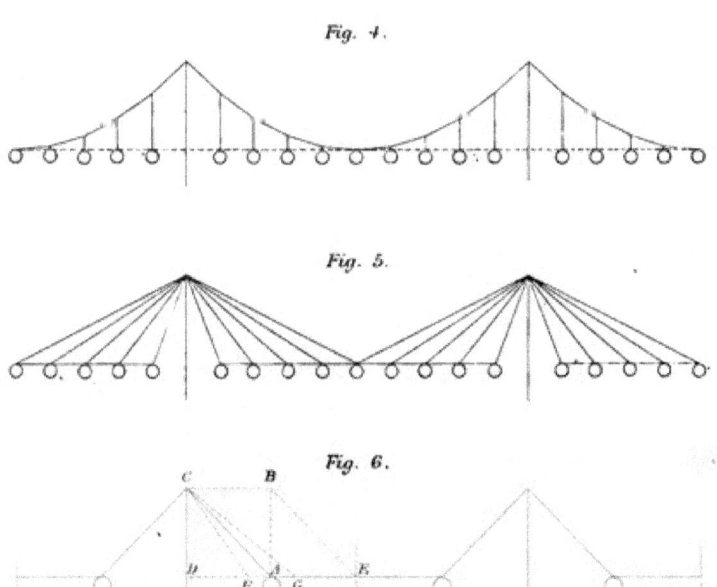

Fig. 4.

Fig. 5.

Fig. 6.

Plate XII

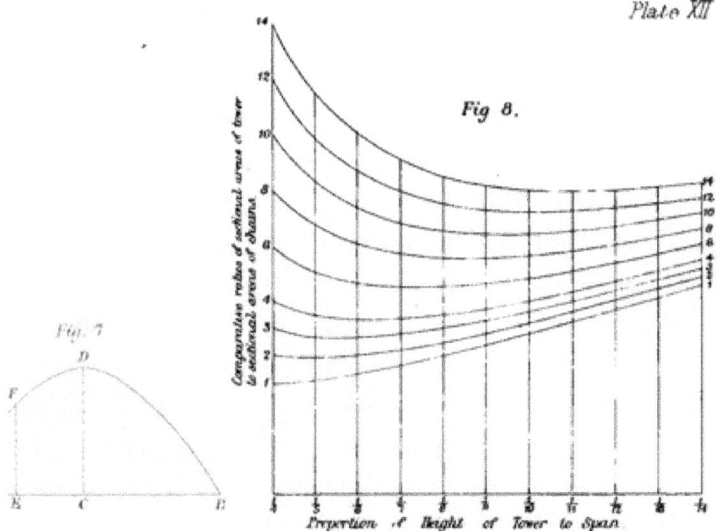

Fig. 7.

Fig. 8.

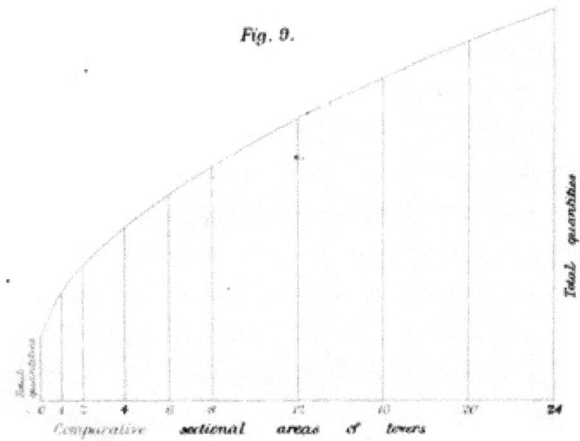

Fig. 9.

Fig. 10.

Plate XIII

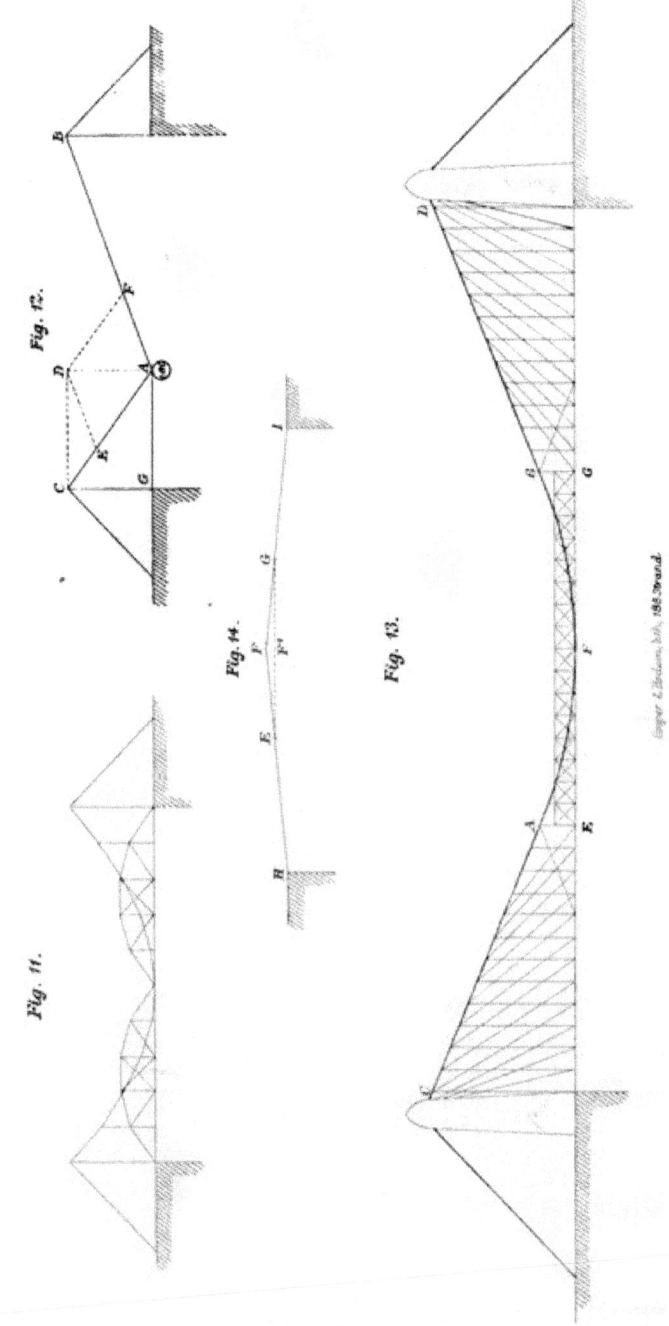

Fig. 12.

Fig. 11.

Fig. 14.

Fig. 13.

www.ingramcontent.com/pod-product-compliance
Lightning Source LLC
Chambersburg PA
CBHW031455160426
43195CB00010BB/985